GCSE (9 – 1): MATHEMATICS HIGHER

Examination Practice Workbook

a Workbook with 578 Exercises & Problems
to ensure success
in the AQA, OCR, EDEXCEL & WJEC Examinations,

by

DR GERASIMOS MASOUROS
ANTHONY NICOLAIDES

PASS PUBLICATIONS: PRIVATE ACADEMIC AND SCIENTIFIC STUDIES LIMITED

PASS PUBLICATIONS
PRIVATE ACADEMIC AND SCIENTIFIC STUDIES LIMITED

First Published in the United Kingdom 2020 by

Private Academic and Scientific Studies Limited

ISBN 9780995742291

FIRST EDITION 2020

PREFACE

We all know that Mathematics is not easy to learn and that it must be read again and again. The first time helps to comprehend the essentials, the second time provides an understanding of the details to some extent and so on.

With such thoughts, we published the «GCSE (9-1) MATHEMATICS HIGHER» a step by step and topic by topic Tutorial Textbook (ISBN 978-1-872684-51-2), which starts from the basics (assuming nothing) and provides all the necessary step by step mathematical processes and theoretical notes that the students must learn. In each Topic, the theory is followed by plenty of fully worked out examples (578 in total), set in progressive difficulty, up to the level of the problems/exercises/questions that all the examination boards would ask.

The reviews we received for this Textbook are really good, the book was sold out and it was reprinted. Now, we are following up with this book, a Companion students' Workbook with Questions and Problems and enough space for the students to solve them.

The answers to the Questions and Problems are highlighted in their headings, hence readily available – not hidden in the back pages, because we do not want the students to be disorientated in the complexity of their thoughts, while they have not mastered the necessary mathematical ways of thinking yet. If they need help, they can read the corresponding Worked Examples from our Tutorial Textbook, and if they do not remember the theory or a mathematical rule/formula, they can read it again in the corresponding Topic, or in the concise Appendix at the end of our Tutorial Textbook.

The Tutorial Textbook (ISBN 978-1-872684-51-2) and this Workbook, which start from the basics (assuming nothing), are the perfect combination providing guidance to the students by showing them how and what to READ and PRACTICE, in order to score a high grade in their exams.

For the school teachers and tutors, this Workbook is a good collection of the assignments they would give homework to their students.

The success is not far. It is within reach and it becomes much easier with the guidance of our two books.

January, 2020

TABLE OF CONTENTS

4. GEOMETRY AND MEASURES 129

5. PROBABILITY 223

6. STATISTICS 253

1 NUMBER

The difficulty level of the following exam-style questions and problems is progressively increasing and their full, step-by-step solutions with the required theoretical notes can be found in the book «GCSE (9-1) Mathematics Higher»[1], pp. 1-46. The numbering of the questions has remained the same, so, for example, the full solution of «Exam Practice Question 1.1» can be found in «Worked Example 1.1.» of the above book. With this arrangement, it is simple for the students to read, understand and remember the theory and also to check and correct their work.

Topic 1.1 BIDMAS.

EXAM PRACTICE QUESTION 1.1 (ANSWER: 43)

Find the answer of $3+8\times5$

EXAM PRACTICE QUESTION 1.2 (ANSWER: 33)

Find the answer of $3\times(5\times2+1)$

EXAM PRACTICE QUESTION 1.3 (ANSWER: 36)

Find the answer of $(4^2+2)\times(2^3\div4)$

EXAM PRACTICE QUESTION 1.4 (ANSWER: 32)

Find the answer of $14+7+(5\div5)-3\times4-11+3+(8+7)\times2$

EXAM PRACTICE QUESTION 1.5 (ANSWER: 42)

Find the answer of $14+7+(5\div5)-3\times4-1^1+3+(8+7)\times2$

[1] Dr Gerasimos Masouros & Anthony Nicolaides, «GCSE (9-1) Mathematics Higher», PASS Publications Ltd, London 2018, ISBN 9781872684512

Topic 1.2 FACTORS AND PRIME FACTORS. LCM. HCF.

EXAM PRACTICE QUESTION 1.6 (ANSWERS: (a) $2^5 \times 3$, $2^2 \times 3 \times 5$, (b) 12, (c) 480)

(a) Express the following numbers as products of their prime factors: (i) 96 (ii) 60
(b) Find the highest common factor (HCF) of 60 and 96
(c) Find the lowest common multiple (LCM) of 96 and 60

EXAM PRACTICE QUESTION 1.7 (ANSWERS: (a) 2^5, (b) 3^4, (c) $2^{13} \times 5^{13}$)

Write down the following numbers as a product of their prime factors in index form:
(a) 32 (b) 81 (c) 10 000 000 000 000

EXAM PRACTICE QUESTION 1.8 (ANSWER: 50)

Given that the LCM of two numbers is 100 and their HCF is 10, find one of the numbers, when the other is 20.

EXAM PRACTICE QUESTION 1.9 (ANSWER: 80)

Given that the HCF of two numbers is 10 and the product of their non-common factors is 24, find one of the numbers, when the other is 30.

Topic 1.3 INDICES AND THE RECIPROCAL OF A POSITIVE NUMBER.

EXAM PRACTICE QUESTION 1.10 (ANSWERS: (a) 2^9, (b) 2^2, (c) 2^3, (d) 2^{15})

Simplify the following expressions, that is write each one of them as a single power of 2:

(a) $2^4 \times 2^5$ (b) $\dfrac{2^5}{2^3}$ (c) $\dfrac{2^6 \times 2^3}{2 \times 2^5}$ (d) $\left(2^3\right)^5$

EXAM PRACTICE QUESTION 1.11 (ANSWERS: (a) 3^7, (b) 3^5, (c) 3^3, (d) 3^{14})

Simplify the following expressions:

(a) $3^3 \times 3^4$ (b) $\dfrac{3^7}{3^2}$ (c) $\dfrac{3 \times 3^7}{3^2 \times 3^3}$ (d) $\left(3^7\right)^2$

EXAM PRACTICE QUESTION 1.12 (ANSWERS: (a) 5^5, (b) 5^{-4}, (c) 5^8, (d) 5^{15})

Simplify the following expressions:

(a) $5^2 \times 5^3$ (b) $\dfrac{5^4 \times 5^5}{5^6 \times 5^7}$ (c) $\dfrac{5^8}{5^0}$ (d) $\left(5^3\right)^5$

Exam Practice Question 1.13 (Answer: 3^{12})

Simplify the expression $27^2 \times 3^6$, that is write $27^2 \times 3^6$ as a single power of 3

Exam Practice Question 1.14 (Answer: 2^5)

Simplify the expression $32^3 \div \left(8 \times 2^7\right)$, that is write $32^3 \div \left(8 \times 2^7\right)$ as a single power of 2.

Exam Practice Question 1.15 (Answers: (a) $\frac{1}{125}$, (b) 25, (c) $\frac{1}{5}$)

Work out the value of (a) 5^{-3} (b) $625^{\frac{1}{2}}$ (c) $3\,125^{-\frac{1}{5}}$

Exam Practice Question 1.16 (Answers: (a) $\frac{1}{5^4}$, (b) $\frac{1}{5^9}$, (c) $\frac{1}{5^{16}}$, (d) $\frac{5^3}{2^3 \times 3^3}$, (e) $\frac{2^2}{3^2}$)

Write the following as indices of prime numbers

(a) 25^{-2} (b) 125^{-3} (c) 625^{-4} (d) $\left(\frac{36}{25}\right)^{-\frac{3}{2}}$ (e) $\left(\frac{81}{16}\right)^{-\frac{1}{2}}$

EXAM PRACTICE QUESTION 1.17 (ANSWER: $5^{\frac{3}{2}}$)

Work out the value $\frac{125}{\sqrt{5}} \times 25^{-\frac{1}{2}}$

EXAM PRACTICE QUESTION 1.18 (ANSWERS: (a) 2^8, (b) 3^7, (c) 11, (d) 13^2)

Simplify the following expressions:

(a) $2^3 \times 2^5$ (b) $\frac{3^9}{3^2}$ (c) $11^7 \div 11^6$ (d) $13^5 \div 13^3$

EXAM PRACTICE QUESTION 1.19 (ANSWERS: (a) 5^{-3}, (b) 17^3, (c) 5^{-1})

Simplify the following expressions:

(a) $5^2 \times 5^{-5}$ (b) $17^4 \div 17$ (c) $\left(5^3\right)^{-\frac{1}{3}}$

EXAM PRACTICE QUESTION 1.20 (ANSWERS: (a) 3^{-3}, (b) 10^{-5}, (c) 2^{-3}, (d) 2^{11})

Work out the value of each of the following:

(a) $81^{-\frac{3}{4}}$ (b) $100^{-\frac{5}{2}}$ (c) $256^{-\frac{3}{8}}$ (d) $(0.5)^{-5} \times 2^6$

Topic 1.4 FRACTIONS.

EXAM PRACTICE QUESTION 1.21 (ANSWER: $3\frac{9}{10}$)

Find the value of $1\frac{1}{2}$ plus $2\frac{2}{5}$

EXAM PRACTICE QUESTION 1.22 (ANSWER: $10\frac{8}{63}$)

Add the mixed fractions $4\frac{4}{7}$ and $5\frac{5}{9}$

EXAM PRACTICE QUESTION 1.23 (ANSWER: $3\frac{1}{35}$)

Work out $5\frac{3}{5}-2\frac{4}{7}$

EXAM PRACTICE QUESTION 1.24 (ANSWER: $1\frac{1}{20}$)

Subtract $2\frac{1}{5}$ from $3\frac{1}{4}$

EXAM PRACTICE QUESTION 1.25 (ANSWER: $7\frac{5}{7}$)

Find the product of the mixed fractions $3\frac{3}{7}$ and $2\frac{1}{4}$

Exam Practice Question 1.26 (Answer:10)

Determine the product $2\frac{2}{3}\times 3\frac{3}{4}$

Exam Practice Question 1.27 (Answer: $1\frac{9}{10}$)

Determine the quotient $4\frac{3}{4}\div 2\frac{1}{2}$

Exam Practice Question 1.28 (Answer: $4\frac{1}{2}$)

Find $15\div 3\frac{1}{3}$

Exam Practice Question 1.29 (Answer: $6\frac{7}{8}$)

Find $2\frac{1}{2}+1\frac{3}{4}\times 2\frac{1}{2}$

Exam Practice Question 1.30 (Answer: $1\frac{1}{6}$)

Determine $\left(1\frac{3}{4}\right)^2\div 2\frac{5}{8}$

Topic 1.5 DECIMALS.

Exam Practice Question 1.31 (Answer:16.817)

Work out $4.32+0.007+1.5+10.99$

Exam Practice Question 1.32 (Answer:4.688)

Subtract 9.392 from 14.08

Exam Practice Question 1.33 (Answer:0.14445)

Evaluate the product 0.45×0.321

Exam Practice Question 1.34 (Answer:0.0603)

Find the product 0.201×0.3

Exam Practice Question 1.35 (Answer:4)

Find the quotient of 0.48 divided by 0.12

Exam Practice Question 1.36 (Answer:1.1)

Determine $0.121\div 0.11$

Exam Practice Question 1.37 (Answer:40)

Find the value of 0.16 divided by 0.004

Exam Practice Question 1.38 (Answer: $\frac{7}{9}$)

Express $0.\dot{7}$ as $\frac{N}{D}$ where N and D are integers.

Exam Practice Question 1.39 (Answer: $\frac{4}{33}$)

Write $0.\dot{1}\dot{2}$ as a fraction.

Exam Practice Question 1.40 (Answer: $\frac{323}{99}$)

Write $3.\dot{2}\dot{5}$ as a fraction

Exam Practice Question 1.41 (Answer: $\frac{374}{333}$)

Write $1.\dot{1}\dot{2}\dot{3}$ as a fraction.

Exam Practice Question 1.42 (Answer: $\frac{7}{45}$)

Write $0.1\dot{5}$ as a fraction.

Exam Practice Question 1.43 (Answer: $5\frac{857}{900}$)

Write $5.95\dot{2}$ as a fraction.

Exam Practice Question 1.44 (Answer: $\frac{38}{125}$)

Change 0.304 to a fraction.

Exam Practice Question 1.45 (Answer: $1\frac{29}{50}$)

Express 1.58 into a mixed fraction.

Topic 1.6 RATIONAL AND IRRATIONAL NUMBERS (SURDS).

EXAM PRACTICE QUESTION 1.46 (ANSWER: $a = 2 + \pi$, $b = 3 - \pi$)

Find two different irrational numbers a and b such that $a + b$ is rational.

EXAM PRACTICE QUESTION 1.47 (ANSWER: $a = 2 + \pi$, $b = 6 - \pi$)

Find two irrational numbers a and b, such that the perimeter of a rectangle with sides a and b is rational.

EXAM PRACTICE QUESTION 1.48 (ANSWER: $a = 2\sqrt{2}$, $b = \frac{11}{\sqrt{2}}$)

The area of a rectangle with sides a and b is rational although a and b are irrational. Find a pair of values for a and b.

EXAM PRACTICE QUESTION 1.49 (ANSWER: $\frac{3\sqrt{2}}{2}$)

Rationalise the denominator of $\frac{3}{\sqrt{2}}$

Exam Practice Question 1.50 (Answer: $\frac{3+\sqrt{3}}{3}$)

Rationalise the denominator of $\frac{2}{3-\sqrt{3}}$

Exam Practice Question 1.51 (Answers: $3\sqrt{5}$, $6\sqrt{2}$, $\frac{5}{4}\sqrt{\frac{5}{2}}$)

Simplify the following surds:

(i) $\sqrt{45}$ (ii) $\sqrt{72}$ (iii) $\sqrt{\frac{125}{32}}$

Topic 1.7 ROUNDING TO A DECIMAL PLACE. SIGNIFICANT FIGURES.

EXAM PRACTICE QUESTION 1.52 (ANSWER:2.98, 4.594, 10.0)

(a) Write down the number 2.98325 to 2 decimal places.
(b) Write down the number 4.593742 to 3 decimal places.
(c) Round 9.999 to 1 decimal place.

EXAM PRACTICE QUESTION 1.53 (ANSWER: 5)

How many significant figures do we have in the number 24 735?

EXAM PRACTICE QUESTION 1.54 (ANSWER: 3)

How many significant figures do we have in the number 0.00375?

EXAM PRACTICE QUESTION 1.55 (ANSWER: 4)

How many significant figures do we have in the number 4 034?

EXAM PRACTICE QUESTION 1.56 (ANSWER: 10.4, 479 000, 0.390)

Round the following numbers to 3 s.f. (significant figures)
(a) 10.438 (b) 478 556 (c) 0.38950

EXAM PRACTICE QUESTION 1.57 (ANSWER: 5 000. = 5.000×10^3)

Round 5 000.2 to four significant figures.

EXAM PRACTICE QUESTION 1.58 (ANSWER: 2.50×10^3)

Write 2 500 with THREE significant figures.

Topic 1.8 APPROXIMATION. ESTIMATION. UPPER AND LOWER BOUNDS.

EXAM PRACTICE QUESTION 1.59 **(ANSWERS: Upper / Lower (a) 1 050 / 950, (b) 505 / 495, (c) 50.5 / 49.5, (d) 70.5 / 69.5, (e) 21 / 19)**

Calculate the Upper and Lower bounds for the following quantities:

(a) 1 000 m to the nearest 100 m

(b) 500 m to the nearest 10 m

(c) 50 cm to the nearest cm (1 cm)

(d) 70 mm to the nearest 1 mm

(e) 20 kg to the nearest kg (2 kg)

EXAM PRACTICE QUESTION 1.60 **(ANSWER: Upper / Lower = 62.5 Kg / 61.5 Kg)**

The mass of a boy is 62 kg correct to the nearest kg. What are Upper and Lower bounds of the mass?

EXAM PRACTICE QUESTION 1.61 **(ANSWER: Upper / Lower = 9.5 s / 10.5 s)**

The winning time of a race was measured as 10 seconds correct to 2 s.f. What are the Upper and Lower bounds?

EXAM PRACTICE QUESTION 1.62 (ANSWER: Upper / Lower = 215.25 m^2 / 185.25 m^2)

Assume that a rectangular garden is measured 10 m by 20 m to the nearest meter. Determine the Upper and Lower bounds of the area of this garden.

EXAM PRACTICE QUESTION 1.63 (ANSWER: Upper / Lower = 12.0 $\frac{g}{cm^3}$ / 11.4 $\frac{g}{cm^3}$)

The mass of a block is 410 g to the nearest 10 g. The volume of the same block is 35 cm^3 to the nearest cm^3. What are the Upper and Lower bounds for the density?

Topic 1.9 STANDARD FORM $A\times10^n$, WHERE $1\le A<10$ AND n IS AN INTEGER.

EXAM PRACTICE QUESTION 1.64 (ANSWERS: the standard form would be: (i) 1.2×10^4, (ii) 5×10^1, (iii) 7×10^0, (iv) 1.3×10^{-5})

Explain why these numbers are not written in standard form.

(i) 12×10^3 (ii) 5×10 (iii) 7 (iv) $1.3\div10^5$

EXAM PRACTICE QUESTION 1.65 (ANSWERS: (i) 2.4×10^4, (ii) 4.0103×10^5, (iii) 5.03×10^{-4})

Write the following numbers in standard form:

(i) 24 000 (ii) 401 030 (iii) 0.000503

EXAM PRACTICE QUESTION 1.66 (ANSWER: 1.2×10^8)

Calculate $\left(4\times10^5\right)\times\left(3\times10^2\right)$

EXAM PRACTICE QUESTION 1.67 (ANSWER: 2×10^6)

Calculate $\left(6\times10^5\right)\div\left(3\times10^{-1}\right)$

EXAM PRACTICE QUESTION 1.68 (ANSWERS: (i) 1.6×10^{16}, (ii) 2×10^{4})

Calculate the following by writing your answer in standard form:

(i) $\left(2\times10^{8}\right)\times\left(8\times10^{7}\right)$ (ii) $\left(3\times10^{2}\right)\div\left(1.5\times10^{-2}\right)$

EXAM PRACTICE QUESTION 1.69 (ANSWERS: (i) 357 000, (ii) 0.00305, (iii) 7.5)

Write the following standard form as ordinary numbers:

(i) 3.57×10^{5} (ii) 3.05×10^{-3} (iii) 7.5×10^{0}

EXAM PRACTICE QUESTION 1.70 (ANSWERS: (i) 7.986×10^{3}, (ii) 7.59×10^{2}, (iii) 3.65×10^{-3})

Express the following ordinary numbers to the standard form:

(i) 7 986 (ii) 759 (iii) 0.00365

EXAM PRACTICE QUESTION 1.71

(ANSWERS: (i) 2.67×10^{2}, (ii) 4.747×10^{4}, (iii) 9.24×10^{6}, (iv) 3.875×10^{5})

Find in standard form the following:

(i) The product of 3.56×10^{3} by 7.5×10^{-2}

(ii) The quotient of 3.56×10^{3} by 7.5×10^{-2}

(iii) The sum of 5.4×10^{5} and 8.7×10^{6}

(iv) The difference between 4.25×10^{5} and 3.75×10^{4}

Topic 1.10 FACTORIALS.

EXAM PRACTICE QUESTION 1.72 (ANSWERS: 2, 6, 24, 120)

Calculate 2! , 3! , 4! and 5!

EXAM PRACTICE QUESTION 1.73

Make a table of factorials from 0 to 10 and observe their rapid increase. Use a calculator.

Topic 1.11 BASIC COUNTING RULES.

EXAM PRACTICE QUESTION 1.74 (ANSWER: 25)

In a class there are 10 boys and 15 girls. 13 of these students wear white shirts and 12 wear other colours. 5 of the students of the class have long hair and 20 have short hair. 2 are less than 1.50 m high and 23 are more than or equal to 1.50 m high. Count all the students of the class and show that it does not really matter how we group the students.

EXAM PRACTICE QUESTION 1.75 (ANSWER: 70)

Assume there are two cupboards; One contains 10 shirts, the other 7 pairs of trousers. In how many ways can you combine all the shirts with all the trousers?

EXAM PRACTICE QUESTION 1.76 (ANSWERS: (i) 280, (ii) £63.40)

A restaurant menu has 5 starters, 8 main dishes and 7 desserts. A customer looks at the menu and observes that he or she has a deal.

a. A starter, a main and a dessert £15.85

b. A starter and a main £12.90

c. A main and a dessert £11.90

Children's menus are reduced by 50%.

i. Work out the different ways of choosing a meal at this restaurant.

ii. A family of 3 adults and 2 children eat at this restaurant and choose (a), what is the total bill?

EXAM PRACTICE QUESTION 1.77 (ANSWERS: (i) 1, (ii) 3, (iii) 3)

Find the number of combinations of 3 different items

i. taken three at a time

ii. taken two at a time

iii. taken one at a time.

Exam Practice Question 1.78 (Answers: (i) 840, (ii) 6 720, (iii) 720)

Write the number of permutations of

i. 7 objects taken 4 at a time

ii. 8 objects taken 5 at a time

iii. 10 objects taken 3 at a time
and evaluate the outcome.

Exam Practice Question 1.79 (Answer: 1 420 494 075)

For 56 matches of football, determine the number of combinations required so that 8 matches are drawn.

Exam Practice Question 1.80 (Answer: 1 048 576

A test consists of 10 multiple choice questions. Each question has 4 choices. How many different possible answers are there?

Exam Practice Question 1.81 (Answer: 9 000 000 000)

A telephone number is given by a sequence of ten digits.
How many different telephone numbers are there if the first digit cannot be 0?

EXAM PRACTICE QUESTION 1.82 (ANSWERS: (a) 10 000, (b) 256, (c) 625, (d) 256)

John has a briefcase with a Permutation lock.

(a) John has to set each part of the lock to a digit between 0 and 9 inclusive. One possible way to this is:

1	2	3	4

In how many different ways can John do this?

(b) John decides that the four digits between 0 and 9 have to be prime numbers. One possible way to this is:

2	3	5	7

In how many different ways can John do this?

(c) John decides to use odd numbers. In how many different ways can John do this?

(d) John decides to use even numbers, excluding zero. In how many different ways can John do this?

EXAM PRACTICE QUESTION 1.83 (ANSWER: 107)

A committee is to be chosen from among 5 mathematicians, 6 physicists and 7 biologists. If the committee is to have two members of different backgrounds, how many such committees can be chosen from the above group of scientists?

EXAM PRACTICE QUESTION 1.84 (ANSWER: YES)

The population of a town is 20 000. If each resident has one, two or three initials, is it true that there must be at least two individuals with the same initials?

2 ALGEBRA

The difficulty level of the following exam-style questions and problems is progressively increasing and their full, step-by-step solutions with the required theoretical notes, diagrams and figures can be found in the book «GCSE (9-1) Mathematics Higher»[1], pp. 47-166. The numbering of the questions has remained the same, so, for example, the full solution of «Exam Practice Question 2.1» can be found in «Worked Example 2.1.» of the above book. With this arrangement, it is simple for the students to read, understand and remember the theory and also to check and correct their work.

Topic 2.1 ALGEBRAIC EXPRESSIONS AND LAWS OF INDICES.

EXAM PRACTICE QUESTION 2.1 (ANSWERS: (a) m^9, (b) n^5, (c) p^4, (d) q^6)

Simplify the following expressions:

(a) $m^4 \times m^5$ (b) $n^3 \times n^2$ (c) $p \times p^3$ (d) $q^2 \times q^4$

EXAM PRACTICE QUESTION 2.2 (ANSWERS: (a) m, (b) n, (c) p^2, (d) q^2)

Simplify the following expressions:

(a) $\frac{m^3}{m^2}$ (b) $\frac{n^4}{n^3}$ (c) $\frac{p^4}{p^2}$ (d) $\frac{q^3}{q}$

---

[1] Dr Gerasimos Masouros & Anthony Nicolaides, «GCSE (9-1) Mathematics Higher», PASS Publications Ltd, London 2018, ISBN 9781872684512

Exam Practice Question 2.3 (Answers: (a) m^6, (b) n^6, (c) p^{20}, (d) q^{20})

Simplify the following expressions:

(a) $(m^2)^3$ (b) $(n^3)^2$ (c) $(p^4)^5$ (d) $(q^5)^4$

Exam Practice Question 2.4 (Answers: (a) m^9, (b) n^5, (c) p^{12}, (d) $\frac{m^9}{n^5}$)

Simplify the following expressions:

(a) $m^7 \times m^2$ (b) $n^8 \div n^3$ (c) $(p^3)^4$ (d) $\frac{m^7 \times m^2}{n^8 \div n^3}$

Exam Practice Question 2.5 (Answers: (a) a^6, (b) b^6, (c) x, (d) z)

Simplify the following expressions:

(a) $a \times a^5$ (b) $b^4 \div b \times b^3$ (c) $\frac{x^2 \times x^3}{x^4}$ (d) $\frac{z^5}{z \times z^3}$

EXAM PRACTICE QUESTION 2.6 (ANSWERS: (a) $12a^3b^4$, (b) $7c^3$, (c) $27d^3e^6$, (d) $25f^6g^8$)

Simplify the following expressions:

(a) $3a^2b^3 \times 4ab$ (b) $\dfrac{35c^3c^4}{5cc^3}$ (c) $\left(3d \times e^2\right)^3$ (d) $\left(f^3 \times 5g^4\right)^2$

EXAM PRACTICE QUESTION 2.7 (ANSWERS: (a) $14x^5y^5$, (b) $27x^2$, (c) $9x^4y^6$, (d) $216x^9y^3$)

Simplify the following expressions:

(a) $7x^2y^3 \times 2x^3y^2$ (b) $\dfrac{81}{3}x^{3+5-(2+4)}$ (c) $\left(3x^2y^3\right)^2$ (d) $\left(2x \times 3x^2y\right)^3$

EXAM PRACTICE QUESTION 2.8 (ANSWERS: (a) $\dfrac{5ab}{3}$, (b) $\dfrac{16c^4}{b^2}$)

Simplify the following expressions:

(a) $\left(\dfrac{3ab}{5a^2b^2}\right)^{-1}$ (b) $\left(\dfrac{b^2c^3}{4bc^5}\right)^{-2}$

EXAM PRACTICE QUESTION 2.9 **(ANSWERS: (a) $\frac{5|a|}{6}$, (b) $\frac{3a^2}{4x}$, (c) $\frac{y^2}{4|x|}$, (d) $\frac{16y^2}{5z}$, (e) $\frac{3}{2}x^{\frac{1}{3}}y^{\frac{1}{3}}$, (f) $\frac{5}{8}x^{\frac{1}{2}}y^{\frac{1}{2}}$)**

Simplify the following expressions:

(a) $\left(\frac{25a^2}{36}\right)^{\frac{1}{2}}$ (b) $\left(\frac{27a^6}{64x^3}\right)^{\frac{1}{3}}$ (c) $\left(\frac{144x^2}{9y^4}\right)^{-\frac{1}{2}}$

(d) $\left(\frac{125z^3}{216y^6}\right)^{-\frac{1}{3}}$ (e) $\left(\frac{8xy}{27x^2y^2}\right)^{-\frac{1}{3}}$ (f) $\left(\frac{64xy^2}{25x^2y^3}\right)^{-\frac{1}{2}}$

Topic 2.2 SUBSTITUTIONS.

EXAM PRACTICE QUESTION 2.10 (ANSWER: - 27)

If $a=-3$, $b=2$ calculate: $a^2(ab-a)$.

EXAM PRACTICE QUESTION 2.11 (ANSWER: - 1 600)

If $x=5$, $y=0$, $z=-2$, find: $x^2\left(yz-(x+3)^2\right)$.

EXAM PRACTICE QUESTION 2.12 (ANSWER: 20 m/s)

In kinematics, we use the formula $v^2=u^2+2as$, where v is the final velocity, u is the initial velocity, a is the acceleration and s is the displacement. Determine v, when $u=10$ m/s, $a=2$ m/s^2 and $s=75$ m.

Topic 2.3 EXPANDING BRACKETS.

EXAM PRACTICE QUESTION 2.13 (ANSWER: $15x^2-16x-21$)

Expand and simplify: $5x(3x-2)-3(2x+7)$.

EXAM PRACTICE QUESTION 2.14 (ANSWERS: $10x^2+13x-3$, $x=-\frac{3}{2}$ and $x=\frac{1}{5}$)

Expand and simplify: $(2x+3)(5x-1)$. Hence solve $10x^2+13x-3=0$

EXAM PRACTICE QUESTION 2.15 (ANSWER: $x=-1$)

Solve $5(2x-1)-3(x-5)=3$

Topic 2.4 LINEAR FACTORISATION.

EXAM PRACTICE QUESTION 2.16 (ANSWERS: (a) $3(x-3)$, (b) $x(x-6)$, (c) $5(a+3)$, (d) $a(a-1)$, (e) $mn(1-mn)$, (f) $7y(x+4)$, (g) $4ab(b-2a)$, (h) $2ax^2(1+3ax)$)

Factorise the following expressions:

(a) $3x-9$ (b) x^2-6x (c) $5a+15$ (d) a^2-a (e) $mn-m^2n^2$

(f) $7xy+28y$ (g) $4ab^2-8a^2b$ (h) $2ax^2+6a^2x^3$

EXAM PRACTICE QUESTION 2.17 (ANSWERS: (a) $3x(x+1)(x-1)$, (b) $3b(2ac+1)(2ac-1)$, (c) $(x-5y)(2x+3y)$)

Factorise the expressions:

(a) $3x^3-3x$ (b) $12a^2bc^2-3b$ (c) $2x^2-7xy-15y^2$

Remember that $\left(x^2-a^2\right)=\left(x+a\right)\left(x-a\right)$

Topic 2.5 SIMPLIFYING ALGEBRAIC FRACTIONS.

EXAM PRACTICE QUESTION 2.18 **(ANSWER: $\frac{4(x+2)}{(x+1)(x+5)}$)**

For $x \neq -1,\ -5$ simplify $\frac{1}{x+1}+\frac{3}{x+5}$

EXAM PRACTICE QUESTION 2.19 **(ANSWER: $\frac{2(x+2)}{(x-1)}$)**

For $x \neq 1,\ 2, -2$ simplify: $\frac{x^2-4}{3x+6} \times \frac{6x+12}{x^2-3x+2}$.

EXAM PRACTICE QUESTION 2.20 **(ANSWER: $\frac{3x(x-1)}{4}$)**

For $x \neq -1, \frac{1}{2}$ simplify: $\frac{x^3-x}{4x-2} \div \frac{2x+2}{6x-3}$.

Exam Practice Question 2.21 (Answer: $x = 1$ and $x = -3$)

For $x \neq -5, -1$ solve: $\dfrac{1}{x+1} + \dfrac{3}{x+5} = 1$.

Exam Practice Question 2.22 (Answer: $\dfrac{4x^2 + 15x + 4}{(2x-1)(2x+1)}$)

For $x \neq \pm\dfrac{1}{2}$, simplify $\dfrac{2x+4}{2x-1} + \dfrac{5x}{4x^2-1}$.

EXAM PRACTICE QUESTION 2.23 (ANSWER: $y = -2$)

Solve for y, given that $\dfrac{y+5}{2y+7} = \dfrac{y-3}{2y-1}$, $y \neq -\dfrac{7}{2}, \dfrac{1}{2}$

EXAM PRACTICE QUESTION 2.24 (ANSWER: $y = 6.06$ and $y = -0.06$ correct to 2 d.p.)

Solve for y, giving your answers correct to 2 decimal places: $\dfrac{5y-1}{y+3} = \dfrac{2y+1}{y-2}$, $y \neq -3,\ 2$.

Topic 2.6 IDENTITIES AND EQUATIONS.

Exam Practice Question 2.25 (Answer: $a = 3,\ b = 5$)

Find the values of a and b, for the identity $3x + 5y \equiv ax + by$.

Exam Practice Question 2.26 (Answers: (i) $x^2 + 2xy + y^2$, (ii) $a^2 + 2ab + b^2$)

Multiply out the brackets and simplify

(i) $(x + y)(x + y)$ (ii) $(a + b)(a + b)$

Observe that the results are identities and express these as identities.

Exam Practice Question 2.27 (Answers: (i) $x^2 - 1$, (ii) 1, (iii) $16a^2 - 9$)

(a) Multiply out the brackets

(i) $(x-1)(x+1)$

(ii) $(\sqrt{3}-\sqrt{2})(\sqrt{3}+\sqrt{2})$

(iii) $(4a+3)(4a-3)$ and simplify.

(b) Observe that the relations are identities. Give two simple examples.

Topic 2.7 LINEAR EQUATIONS AND THEIR SOLUTIONS.

EXAM PRACTICE QUESTION 2.28 (ANSWER: $x = \frac{5}{3}$)

In the equation of the straight line, $y = ax + b$, if $y = 4$, $a = 3$ and $b = -1$, then the linear equation is $4 = 3x - 1$ or $3x - 1 = 4$. Solve it for x.

EXAM PRACTICE QUESTION 2.29 (ANSWER: $x = 5$)

Solve the equation $7x - 5 = 30$

EXAM PRACTICE QUESTION 2.30 (ANSWER: $x = \frac{1}{2}$)

Solve the equation $4x + 3 = 5$

EXAM PRACTICE QUESTION 2.31 (ANSWER: $x = 8$)

Solve the linear equation $4(x - 3) = 20$

EXAM PRACTICE QUESTION 2.32 (ANSWER: $x = 8$)

Solve $\frac{2x - 1}{3} = 5$

Exam Practice Question 2.33 (Answer: $x=\frac{14}{5}=2\frac{4}{5}$)

Solve $\frac{2}{3}(5x+1)=10$

Exam Practice Question 2.34 (Answer: $x=-\frac{2}{3}$)

Solve $3(x-2)=-8$

Exam Practice Question 2.35 (Answer: $x=\frac{1}{4}$)

Solve $5\left(2x+\frac{1}{2}\right)=5$

Exam Practice Question 2.36 (Answer: $x=-4$)

Solve $5x+3=2x-9$

EXAM PRACTICE QUESTION 2.37 (ANSWER: $x = 2$)

Solve the linear equation $11(x+9)=121$

EXAM PRACTICE QUESTION 2.38 (ANSWER: $x = 169$)

Solve $\frac{x}{13}-8=5$

EXAM PRACTICE QUESTION 2.39 (ANSWER: $x = 5$)

Solve $3x+7=2x+12$

EXAM PRACTICE QUESTION 2.40 (ANSWER: $x = 10$)

Solve the linear equation $\frac{1}{4}x-\frac{1}{5}x=\frac{1}{2}$

EXAM PRACTICE QUESTION 2.41 (ANSWER: $x=\frac{18}{11}=1\frac{7}{11}$)

Solve the linear equation $\frac{x}{1}+\frac{x}{2}+\frac{x}{3}=3$

EXAM PRACTICE QUESTION 2.42 (ANSWER: $x=\frac{3}{5}$)

Find the value of $x \neq 0$ in the linear equation $\frac{1}{x}+\frac{2}{x}=5$

Topic 2.8 REARRANGING FORMULAE.

EXAM PRACTICE QUESTION 2.43 (ANSWER: $y = \frac{7+b}{a}$)

Make y the subject of $ay - b = 7,\ a \neq 0$

EXAM PRACTICE QUESTION 2.44 (ANSWER: $x = \frac{d}{a} + b$)

Make x the subject of $a(x-b) = d,\ a \neq 0$

EXAM PRACTICE QUESTION 2.45 (ANSWER: $z = \pm\sqrt{\frac{10}{3}}$)

Make z the subject of $3z^2 + 5 = 15$

EXAM PRACTICE QUESTION 2.46 (ANSWER: $v = \pm\sqrt{2as + u^2}$)

Make v the subject of $v^2 - u^2 = 2as$.

EXAM PRACTICE QUESTION 2.47 (ANSWER: $\ell = \frac{T^2 g}{4\pi^2}$)

Make l the subject of $T = 2\pi\sqrt{\frac{l}{g}}$.

EXAM PRACTICE QUESTION 2.48 (ANSWER: $g = \frac{4\pi^2 \ell}{T^2}$)

In the expression of $T = 2\pi\sqrt{\frac{l}{g}}$ make g the subject.

EXAM PRACTICE QUESTION 2.49 (ANSWER: $x = \frac{b-14}{6-a}$)

Make x the subject of the equation $2(3x+7) = ax + b,\ \ a \neq 6$.

EXAM PRACTICE QUESTION 2.50 (ANSWER: $x = \frac{c-b}{a+b}$)

Make x the subject of the equation $ax + b = c - bx$, $a \neq -b$.

EXAM PRACTICE QUESTION 2.51 (ANSWER: $y = -\frac{k}{h}x + k$)

Make y the subject of the equation $\frac{x}{h} + \frac{y}{k} = 1$, $h, k \neq 0$.

Topic 2.9 COORDINATE GEOMETRY. PLOTTING LINEAR GRAPHS WITH THE USE OF A TABLE OF VALUES. DRAWING GRAPHS FROM LINEAR EQUATIONS IN THE FORM $y = mx + c$.

EXAM PRACTICE QUESTION 2.52 **(ANSWER: The lines are parallel, the intercepts are different)**

Make a table of values for $y = 3x - 1$ when the values of x are: $-3, -2, -1, 0, 1, 2, 3$.

Hence plot the points and draw the graph.

Repeat a table of values for the linear equation $y = 3x + 2$ and draw the line. What conclusion did you make?

Topic 2.10 THE GRADIENT OF A STRAIGHT LINE AND GRADIENT AT A POINT *P* OF A CURVE.

EXAM PRACTICE QUESTION 2.53 (ANSWERS: (a) $y = m_1 x$, (b) $y = -\frac{1}{m_1}x$, (c) $AB = \frac{m_1^2+1}{m_1}$)

(a) Write down the equation of a line l_1 passing through the origin.

(b) Write down the equations of the line l_2 which is passing through the origin and is perpendicular to the above l_1.

(c) Given that the points A and B are lying on l_1 and l_2 respectively and that the x coordinate of both equals to 1, find the lengths OA, OB and AB in terms of m_1 and m_2, where O is the origin.

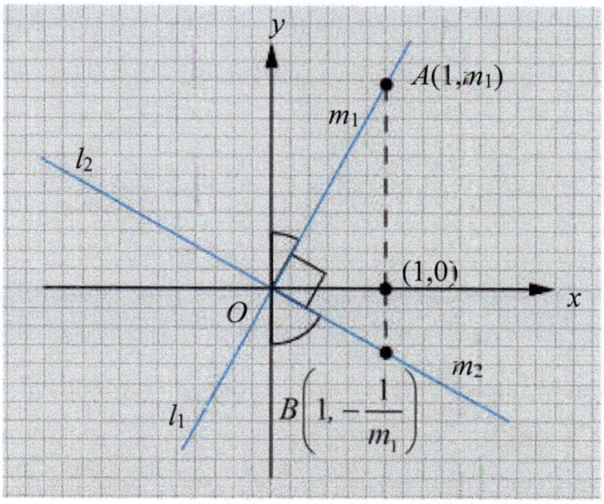

EXAM PRACTICE QUESTION 2.54 (ANSWER: $m_1 \times m_2 = -1$)

Show that the straight lines $y = -2x + 1$ and $y = \frac{1}{2}x + 3$ are perpendicular.

EXAM PRACTICE QUESTION 2.55 (ANSWER: $m_1 \neq m_2$ and $m_1 \times m_2 \neq -1$)

Show that the lines with equations $3x + 5 - y = 0$ and $x - 3 + y = 0$ are neither parallel nor perpendicular.

EXAM PRACTICE QUESTION 2.56 (ANSWER: $m_1 = m_2$, hence the lines are parallel)

Make y the subject for (i) $3x + 4y + 5 = 0$ and (ii) $3x + 4y - 3 = 0$ and show that the lines are parallel.

EXAM PRACTICE QUESTION 2.57 (ANSWER: $m = 1$, and $M(6, 2)$)

Two sets of points $P(7, 3)$ and $Q(5, 1)$ define a straight line PQ. Find the gradient of the line PQ and the coordinates of its midpoint M.

EXAM PRACTICE QUESTION 2.58 (ANSWER: $2\sqrt{2}$)

Find the exact value of the length PQ, where $P(7, 3)$ and $Q(5, 1)$.

EXAM PRACTICE QUESTION 2.59 (ANSWER: $y = x - 4$)

Find the equation of the line PQ, where $P(7, 3)$ and $Q(5, 1)$.

EXAM PRACTICE QUESTION 2.60 (ANSWER: $m = \frac{4}{3}$)

Plot the two sets of points, $A(5,\ 3)$ and $B(2,\ -1)$, on the linear graph paper. Draw the line through these two points and determine its gradient.

EXAM PRACTICE QUESTION 2.61 (ANSWERS: (a) $m = 1$, (b) $y = x + 2$, (c) $(0,\ 2)$ (d) they both lie on the line)

Plot the points $A(4,\ 6)$ and $B(1,\ 3)$ on a linear graph paper and draw the line through these points. Then determine:

(a) The gradient of this line
(b) Find the linear equation which represents the above straight line
(c) Find the intercept on the y axis
(d) Check that the points (2, 4) and (3, 5) lie on this line

EXAM PRACTICE QUESTION 2.62 (ANSWERS: (a) $m=-\frac{3}{2}$, (b) $y=-\frac{3}{2}x+3$)

A straight line is defined by the two points $A(0, 3)$ and $B(2, 0)$.

(a) Find the gradient of the line AB.

(b) Find the equation of the line and hence write down the intercept on the y-axis.

EXAM PRACTICE QUESTION 2.63 (ANSWER: $y=x+3,\ y=-x+3$)

Given the points $P(7, 3)$ and $Q(5, 1)$, determine the equation of a line parallel to PQ which is passing through a point $S(0, 3)$. Also, find the perpendicular line that passes through $S(0, 3)$.

EXAM PRACTICE QUESTION 2.64 (ANSWER: YES, because $m_{AB} \times m_{AC} = -1$)

Plot the points A (4, 2), B (3, 3) and C (5, 3). Is the triangle ABC a right-angled triangle, or equivalently, is one of its angles equal to 90 degrees?

EXAM PRACTICE QUESTION 2.65 **(ANSWERS: gradient of $AE = 5$, of $AD = 4.8$, of $AC = 4.5$, of $AB = 4.2$, We deduce that the tangent at A becomes the gradient of the curve at A)**

Sketch the graph of $y = x^2$ between the values of 2 and 3, and insert the following coordinates of its points, *A*, *B*, *C*, *D* & *E* on the curve:

	A	*B*	*C*	*D*	*E*
x :	2	2.2	2.5	2.8	3
$y = x^2$:	4	4.84	6.25	7.84	9

TABLE 2.10.1

Join the points to give *AE*, *AD*, *AC*, *AB* as the chords. Workout the gradients of the 4 chords, *AB*, *AC*, *AD* and *AE* and observe that the gradient of the chords decreases as the distance of the points *E*, *D*, *C* and *B* form *A* tends to zero.

Draw the tangent at *A*. What do you deduce?

Topic 2.11 QUADRATIC EXPRESSIONS. QUADRATIC EQUATIONS. FACTORISATION. QUADRATIC GRAPHS.

Method 1: Factorisation of a quadratic expression, when $\alpha = 1$.

EXAM PRACTICE QUESTION 2.66 (ANSWERS: (a) $(x+6)(x+4)$, (b) $(x-6)(x-4)$, (c) $(x+12)(x+2)$, (d) $(x-12)(x-2)$, (e) $(x+3)(x+8)$, (f) $(x-3)(x-8)$, (g) $(x+1)(x+24)$, (h) $(x-1)(x-24)$)

Factorise the following quadratic expressions:

(a) $x^2+10x+24$ (b) $x^2-10x+24$ (c) $x^2+14x+24$ (d) $x^2-14x+24$

(e) $x^2+11x+24$ (f) $x^2-11x+24$ (g) $x^2+25x+24$ (h) $x^2-25x+24$

EXAM PRACTICE QUESTION 2.67 **(ANSWERS: (a) $x(x-3)$, $x=0$, $x=3$, (b) $(x+1)(x-8)$, $x=-1$, $x=8$, (c) $(2x-7)(2x+7)$, $x=\frac{7}{2}$, $x=-\frac{7}{2}$)**

Factorise and hence solve the following:

(a) $x^2-3x=0$ (b) $x^2-7x-8=0$ (c) $4x^2-49=0$

Method 2 "Using the Formula"

EXAM PRACTICE QUESTION 2.68 **(ANSWER: $x=0.18$, $x=-1.85$)**

Solve the quadratic equation $3x^2+5x-1=0$, giving the solutions to 2 decimal points. Apply Method 2 "Using the Formula"

EXAM PRACTICE QUESTION 2.69 **(ANSWER: $x=1.45$, $x=-3.45$)**

Solve the quadratic equation $x^2+2x-5=0$. Apply Method 2 "Using the Formula" and give the solutions to 3 s.f.

EXAM PRACTICE QUESTION 2.70 (ANSWER: $y=\frac{1+\sqrt{7}}{3},\ y=\frac{1-\sqrt{7}}{3}$)

Solve the quadratic equation $3y^2-2y-2=0$, giving your answers in surd form.

Method 3: "Completing the Square"

EXAM PRACTICE QUESTION 2.71 (ANSWER: $x=-3,\ x=-5$)

Solve the quadratic equation $x^2+8x+15=0$, using the **completing the square method**.

EXAM PRACTICE QUESTION 2.72 (ANSWER: $a=3,\ \beta=-\frac{7}{6},\ \gamma=-\frac{25}{12},\ (x-2)(3x-1)$)

If $3x^2-7x+2\equiv\alpha\left(x+\beta\right)^2+\gamma$, find the values of α, β and γ and hence factorise.

EXAM PRACTICE QUESTION 2.73 (ANSWERS: (a) $a = -\frac{5}{2}$, $b = \frac{3}{4}$, (b) $a = -\frac{7}{2}$, $b = -\frac{85}{4}$, (c) $a = -2$, $b = -2$, (d) $a = 4$, $b = -21$)

For all values of x, find the values of a and b for the following identities:

(a) $x^2 - 5x + 7 \equiv (x+a)^2 + b$

(b) $x^2 - 7x - 9 \equiv (x+a)^2 + b$

(c) $x^2 - 4x + 2 \equiv (x+a)^2 + b$

(d) $x^2 + 8x - 5 \equiv (x+a)^2 + b$

Method 4: "Splitting the Middle Term"

EXAM PRACTICE QUESTION 2.74 **(ANSWER: $(x+3)(2x-1)$)**

Factorise the quadratic expression $2x^2+5x-3$ using **method 4 "Splitting the Middle Term".**

EXAM PRACTICE QUESTION 2.75 (ANSWERS: (a) $(x-2)(3x-1)$ (b) $(2x-5)(2x-1)$, (c) $(2x+5)(3x-1)$)

Factorise the following quadratic expressions by using **the Splitting of the Middle Term** method.

(a) $3x^2-7x+2$ (b) $4x^2-12x+5$ (c) $6x^2+13x-5$

Sketching the graph of the Quadratic Expression

EXAM PRACTICE QUESTION 2.76 (ANSWER: $y=x^2-5x+4$)

For a quadratic equation $x^2+bx+c=0$, it is given that:

(i) it has two solutions m and n,

(ii) the sum of the two solutions equals to 5 and their product is equal to 4.

Find the equation of the parabola and sketch its graph.

EXAM PRACTICE QUESTION 2.77 (ANSWER: $a = 0.5$, $b = 0$, $c = 0$)

For a quadratic expression $y = ax^2 + bx + c$, it is given that:

(i) the x – axis is tangent to its graph at (0, 0) and

(ii) the y – coordinate of one of its graph's points is 0.5, when $x = 1$.

Find the coefficients a, b and c and sketch its graph.

EXAM PRACTICE QUESTION 2.78 (ANSWER: $y = x^2 - 4x + 3$)

For a quadratic equation $ax^2 + bx + c = 0$, it is given that:

(i) it has two solutions $x = 1$ and $x = 3$, (ii) the y – coordinate of its vertex is – 1

Find the equation of the parabola and sketch its graph.

EXAM PRACTICE QUESTION 2.79 (**ANSWERS: (a) vertex (minimum)** $B\left(\frac{5}{2},\frac{3}{4}\right)$, **$x$-intercepts do not exist, y-intercept: (0,7),**

(b) vertex (minimum) $B\left(\frac{7}{2},\frac{85}{4}\right)$, **$x$-intercepts: (- 1.1,0) and (8.1,0) , y-intercept: (0,- 9),**

(c) vertex (minimum) $B(2,-2)$, **x-intercepts: (0.6,0) and (3.4,0) , y-intercept: (0,2),**

(d) vertex (minimum) $B(-4,-21)$, **x-intercepts: (- 8.6,0) and (0.6,0) , y-intercept: (0,- 5))**

Find the vertex, the y-intercept and the x-intercepts (if any) of each one of the following quadratic expressions:

(a) $f(x) = x^2 - 5x + 7$

(b) $f(x) = x^2 - 7x - 9$

(c) $f(x) = x^2 - 4x + 2$ and

(d) $f(x) = x^2 + 8x - 5$

Hence draw their graphs.

Exam Practice Question 2.80

(Answers: (a) $P(0,18)$ and $B(4,2)$, (b) $P(0,4)$ and $B(-3,-5)$)

(a) Fig. 2.11.1 shows the graph of $f(x) = (x-4)^2 + 2$. Find the coordinates of P and B.

(b) Fig. 2.11.2 shows the graph of $g(x) = (x+3)^2 - 5$. Find the coordinates of P and B.

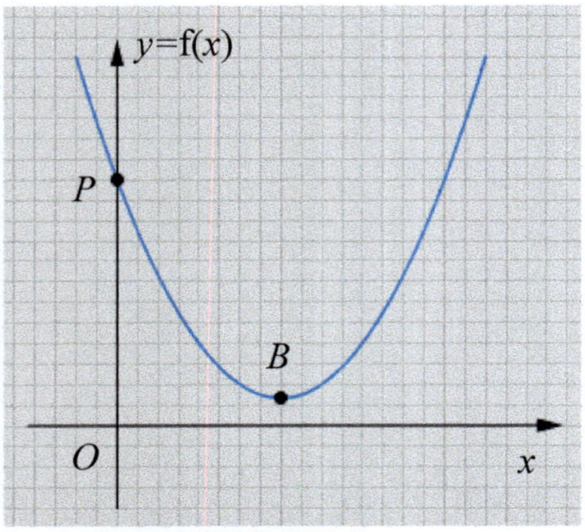

FIG. 2.11.1

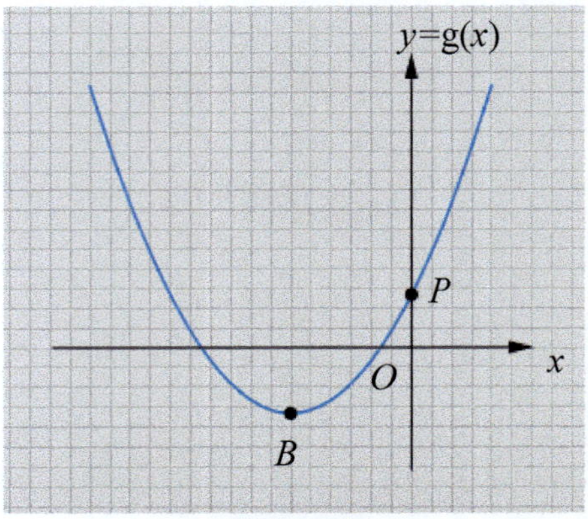

FIG. 2.11.2

EXAM PRACTICE QUESTION 2.81

(ANSWERS: x-intercepts: (- 3.4,0) and (- 0.6,0), y-intercept: (0,2), minimum $B(-2,-2)$,

Find the coordinates of the points of intersections of $f(x) = x^2 + 4x + 2$ with the axes (if any) and also find its minimum value. Hence draw its graph.

EXAM PRACTICE QUESTION 2.82

Make a table of values for the following:

(a) $y = x^2 - x - 2$, for $x = -2, -1, 0, 1, 2, 3$

(b) $y = x^2 - 2x - 3$, for $x = -3, -2, -1, 0, 1, 2, 3, 4$

Hence draw the graphs finding the coordinates of the points of its intersections with the axes.

Note: the detailed solution is presented in page 108 of the book «GCSE (9-1) Mathematics Higher»[2].

[2] Dr Gerasimos Masouros & Anthony Nicolaides, «GCSE (9-1) Mathematics Higher», PASS Publications Ltd, London 2018, ISBN 9781872684512

Topic 2.12 INEQUALITIES.

Linear Inequalities

EXAM PRACTICE QUESTION 2.83 (ANSWER: $x > 3.2$)

Solve the inequality $5x + 1 > 17$, indicating the solution on a number line diagram.

EXAM PRACTICE QUESTION 2.84 (ANSWER: $2 < x \leq 9$)

Solve the inequality $5 < 2x + 1 \leq 19$, indicating the solution on a number line diagram.

EXAM PRACTICE QUESTION 2.85 (ANSWER: $x \geq 3$)

Solve the inequality $2x - 9 \leq 5x - 18$.

EXAM PRACTICE QUESTION 2.86 (ANSWER: $x \leq 3$)

Solve the inequality $2x - 5 \leq 10 - 3x$.

Quadratic Inequalities

EXAM PRACTICE QUESTION 2.87 (ANSWER: $-5 < x < 2$)

Solve $x^2 + 3x < 10$, indicating the solution on a diagram.

EXAM PRACTICE QUESTION 2.88 (ANSWER: $x \geq 4$ or $x \leq -4$)

Solve $x^2 \geq 16$, indicating the solution on a diagram.

EXAM PRACTICE QUESTION 2.89 (ANSWER: $-\frac{1}{2} \le x \le \frac{2}{3}$)

Solve $30x^2 \le 5x + 10$, indicating the solution on a diagram.

EXAM PRACTICE QUESTION 2.90 **(ANSWERS: (a) $x > 2$ or $x < 1$, (b) $1 \le x \le 2$)**

Solve Graphically the inequalities (a) $(x-1)(x-2) > 0$ and (b) $(x-1)(x-2) \le 0$.

Topic 2.13 KINEMATICS OF A PARTICLE – TRANSLATIONS OF SIMPLE SITUATIONS OR PROCEDURES INTO ALGEBRAIC EXPRESSIONS OR FORMULAE.

EXAM PRACTICE QUESTION 2.91

(ANSWERS: (i) 11.11 km, (ii) when the bomb reaches the target, the airplane will be exactly above it)

An airplane (bomber) is cruising above the ground at 800 km/h and its height is $h = 12\ 500$ m and intends to strike a target on the ground from this height, as shown in the following figure.

(i) How far from the target the bomb must be released in order to strike it?

(ii) If the airplane keeps flying at the same direction with the same speed, where will it be when the bomb strikes the target?

(Take $g = 10\ \text{m/s}^2$, consider no friction with the air, no wind blowing and remember 1 h = 3 600 s).

EXAM PRACTICE QUESTION 2.92

(ANSWER: 2 s)

A particle is fired upwards from the roof of a 6 floor-building of height $h = 18$ m, with an initial velocity, $u = 1$ m/s. After how long the particle reaches the ground? (Take $g = 10\ \text{m/s}^2$).

Exam Practice Question 2.93 (Answers: (i) 1.39 m/s^2 to 2 d.p., (ii) 278 m)

A train starts from rest at a station A and moves with uniform acceleration. 20 seconds later, the train passes a point B with a speed of 100 km/h.

Determine

(i) the acceleration of the train in m/s^2 and

(ii) the distance in meters from A to B.

Exam Practice Question 2.94 (Answer: $AD = 2\ 224$ m)

A train starts from rest at a station A and moves with uniform acceleration 1.39 m/s^2 for 20 s, until it reaches point B. Then it travels from B to C with a constant speed for 1 minute and then decelerates with uniform deceleration of a=1.39 m/s^2 till it stops at station D.

Draw the speed/time graph and calculate the total distance covered from A to D.

EXAM PRACTICE QUESTION 2.95 (ANSWERS: (a) 280 m, (b) 20 s, (c) 100.8 km/h)

A train $T1$ starts from rest at a station A and moves with uniform acceleration 1.40 m/s^2. At the same time, another train (train $T2$) passes from the station, moving in the same direction as $T1$, on a parallel line with constant speed equal to 50.4 km/h.

(a) How far from the station will the two trains meet?
(b) After how long and
(c) What is the speed of both trains then?

Topic 2.14 EQUATION OF A CIRCLE WITH CENTRE AT THE ORIGIN, EQUATION OF THE TANGENT AT A GIVEN POINT.

EXAM PRACTICE QUESTION ON THE EQUATION OF THE CIRCLE WITH CENTRE AT THE ORIGIN

(ANSWERS: (a) $x^2 + y^2 = 13$ and $12y + 5x = 169$)

In Fig. 2.14.1, find the equation of the circle and hence find the equation of its tangent at $P(5, 12)$

Note: the detailed solution is presented in page 125 of the book «GCSE (9-1) Mathematics Higher»[3].

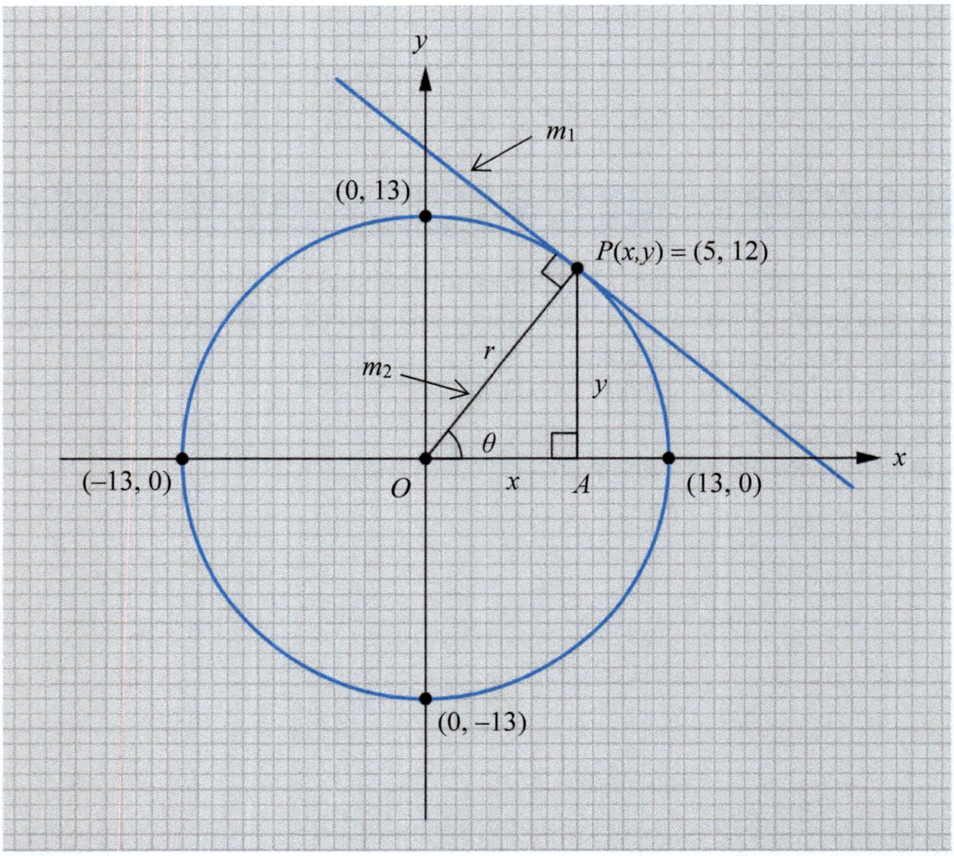

FIG. 2.14.1

[3] Dr Gerasimos Masouros & Anthony Nicolaides, «GCSE (9-1) Mathematics Higher», PASS Publications Ltd, London 2018, ISBN 9781872684512

Topic 2.15 A CUBIC FUNCTION AND ITS GRAPH

EXAM PRACTICE QUESTION 2.96 **(ANSWERS: (a) for $x=-1.5$, $y=5.625$, for $x=-0.5$, y=4.875, for x=1.5, y=0.375, for x=2.5, y=8.625, (c) i: $x=2.8$, ii: $x=-1.7$, or $x=-0.3$, or $x=2.2$, iii: $x=-2.3$, or $x=1$, iv: $x=-2.9$)**

(a) Complete the table of values for $y=x^3-4x+3$

x	–3	–2.5	–2	–1.5	–1	– 0.5	0	0.5	1	1.5	2	2.5	3
y	–12	–2.625	3		6		3	1.125	0		3		18

TABLE 2.15.1

(b) On the grid, draw the graph of $y=x^3-4x+3$

(c) Use the graph to find estimates for

(i) $y=15$ (ii) $y=4$ (iii) $y=0$ (iv) $y=-10$.

Topic 2.16 RECIPROCAL GRAPHS.

Exam Practice Question on the Reciprocal Graphs

(Answers: (i) the x – axis and y – axis, (ii) $y = x$ and $y = -x$)

Draw the reciprocal graphs $y = \frac{1}{x}$ and determine (i) the asymptotes and (ii) the axes of symmetry

Note: the detailed solution is presented in page 125 of the book «GCSE (9-1) Mathematics Higher»[4].

[4] Dr Gerasimos Masouros & Anthony Nicolaides, «GCSE (9-1) Mathematics Higher», PASS Publications Ltd, London 2018, ISBN 9781872684512

Topic 2.17 AN EXPONENTIAL FUNCTION AND ITS GRAPH.

EXAM PRACTICE QUESTION 2.97 (ANSWER: $k=1$, $a=5$)

A curve passes through the points $A(1, 5)$ and $B(3, 125)$.

The curve is represented by the exponential function $y = ka^x$, $a > 0$

Calculate the value of k and the value of a.

EXAM PRACTICE QUESTION 2.98 (ANSWER: $£C \times \left(1+\frac{5}{100}\right)^n$, £3 144.47)

A capital £C is invested at 5% per annum at compound interest for n years. Write down an exponential formula for the total amount received after n years.
If £C is £5 000 what would be the interest earned after 10 years? Draw the graph of this investment.

Exam Practice Question 2.99 (Answer: £9 207,65)

A car is purchased for £35 000 and the depreciation per annum is considered to be 12.5%. What would be the value of the car after 10 years? Draw the graph of the value of the car during the 10 years' period.

Topic 2.18 SIMULTANEOUS LINEAR EQUATIONS.

Algebraic Solution of Simultaneous Linear Equations.

EXAM PRACTICE QUESTION 2.100 (ANSWER: $x = 2$, $y = 1$)

Solve the simultaneous linear equations

$$x + 3y = 5 \quad \ldots (1)$$

$$x - y = 1 \quad \ldots (2)$$

EXAM PRACTICE QUESTION 2.101 (ANSWER: $x = -1$, $y = 2$)

Solve the simultaneous linear equations

$$3x + y = -1 \quad \ldots (1)$$

$$x - 2y = -5 \quad \ldots (2)$$

EXAM PRACTICE QUESTION 2.102 (ANSWER: $x = -5$, $y = -1$)

Solve

$$3x - 5y = -10 \quad \ldots (1)$$

$$-2x + y = 9 \quad \ldots (2)$$

EXAM PRACTICE QUESTION 2.103 (ANSWER: $x=5,\ y=3$)

Solve
$$5x+7y=46 \quad \ldots(1)$$
$$x-y=2 \quad \ldots(2)$$

Graphical Solution of Simultaneous Linear Equations.

EXAM PRACTICE QUESTION 2.104 (ANSWER: $x=2,\ y=1$)

Draw the straight lines:
$$3x-y-5=0 \quad \ldots(1)$$
$$-2x+3y+1=0 \quad \ldots(2)$$

Hence, find the point of intersection and check that each equation is satisfied.

EXAM PRACTICE QUESTION 2.105 (ANSWER: $x = 5$, $y = 7$)

Solve the linear equations

$$x - y = -2 \qquad \dots (1)$$

$$x + y = 12 \qquad \dots (2)$$

simultaneously, using the graphical method. Check your answer with an algebraic method.

Problem solving on simultaneous linear equations

EXAM PRACTICE QUESTION 2.106 **(ANSWER: $s = 16$, $d = 14$, the son is older)**

A family has a daughter and a son. If we add twice the age of the son to the age of the daughter, we find 46 years. While, if we subtract the age of the daughter from twice the age of the son, we find 18 years. Find the age of the son and the age of the daughter. Who is older?

EXAM PRACTICE QUESTION 2.107 **(ANSWER: $g = 7$, $b = 11$)**

Alex invited 12 boys and 10 girls in his birthday party, but 18 friends attended the party. Find how many boys and how many girls did not go to the party, if you know that adding twice the number of boys that went to the party to three times the number of girls that went to the party, you find 43 friends.

Topic 2.19 SIMULTANEOUS SOLUTIONS OF A LINEAR AND A NON LINEAR EQUATION.

EXAM PRACTICE QUESTION 2.108 (ANSWER: $x = -1$, $y = 2$ and $x = 2$, $y = -1$)

Solve the simultaneous equations

$$x^2 + y^2 = 5 \quad \dots (1)$$

$$x + y = 1 \quad \dots (2)$$

Sketch the graphs and indicate the coordinates of the points of intersections.

EXAM PRACTICE QUESTION 2.109 (ANSWER: $x = 3$, $y = 0$ and $x = 8$, $y = 15$)

Solve the simultaneous equations

$$y = x^2 - 8x + 15 \quad \dots (1)$$

$$y + 9 = 3x \quad \dots (2)$$

Sketch the graphs and indicate the coordinates of the points of intersections.

Exam Practice Question 2.110 (Answer: $x = -3$, $y = 0$ and $x = 5$, $y = 8$)

Solve the simultaneous equations

$$y - x - 3 = 0 \qquad \dots (1)$$

$$y = x^2 - x - 12 \qquad \dots (2)$$

Hence sketch the graphs showing the points of intersections.

EXAM PRACTICE QUESTION 2.111 (ANSWER: there is no solution to verify the given questions simultaneously)

For the two Non-Linear Equations shown below:

$$y = \frac{k}{x} \qquad \ldots (1)$$

$$x^2 + y^2 = k^2 \qquad \ldots (2)$$

draw the corresponding graphs on the same grid, for $k = \pm\, 9$. Give the solutions.

Topic 2.20 AREA UNDER CURVES.

EXAM PRACTICE QUESTION 2.112 (ANSWER: 200 km)

A vehicle is traveling at a constant speed of 100 km/hour.
Draw the graph of speed against time and calculate the distance covered in 2 hours.

EXAM PRACTICE QUESTION 2.113 (ANSWER: 4.5 m)

A vehicle starts from rest and accelerates with 1 m/s^2. Draw the graph of speed against time and calculate the distance covered in 3 seconds.

EXAM PRACTICE QUESTION 2.114 (ANSWER: 10.5 m)

A vehicle passes in front of a point A with speed 2 m/s, while accelerating with 1 m/s^2. Draw the graph of speed against time, and calculate the distance covered within the next 3 seconds.

Exam Practice Question 2.115 (Answers: (i) 20 m, (ii) 30 m/s)

A ball is thrown upwards with initial speed $u = 20$ m/s form the roof of a building which is 25 m high. Draw the graph of speed against time, and calculate:

(i) the distance covered when the ball reaches the highest point of its trajectory and

(ii) the ball's velocity, when it reaches the ground

(take $g = 10$ m/s^2).

Exam Practice Question 2.116 (Answer: 30 joules)

A force of 3 N is applied on a particle for as long as the particle's displacement equals to 10 m. Given that the force is parallel to the displacement calculate the work done.

Exam Practice Question 2.117 (Answer: 0.005 joules)

Hooke's law states that the extension of a uniform wire (an elastic material) is directly proportional to the applied load provided that the limit of proportionality has not being exceeded.
From the graph of force against extension in the following figure, calculate the energy stored in an elastic material when it is extended by 0.001 m, or 1 mm under the force of 10 N.

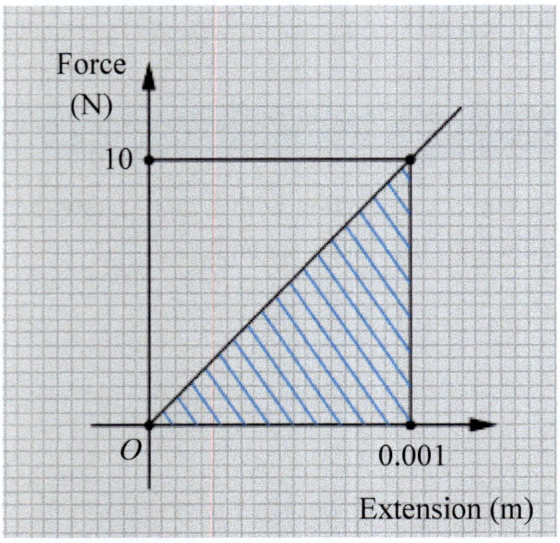

FIG. 2.20.1

Topic 2.21 MAPPING DIAGRAMS, FUNCTION, INVERSE FUNCTION AND COMPOSITE FUNCTION.

EXAM PRACTICE QUESTION 2.118 **(ANSWER: $f^{-1}(x) = x + 1$, it is a function)**

i. For the function $f(x) = x - 1$, draw the mapping diagram for $D = \{1, 2, 3\}$.

ii. Find the inverse relation $f^{-1}(x)$ and draw the mapping diagram. Is this relation a function?

Exam Practice Question 2.119 (Answer: $f^{-1}(x) = x + 1$)

Find the inverse function of $f(x) = x - 1$ and draw the graphs of $f(x)$ and $f^{-1}(x)$.

Exam Practice Question 2.120 (Answer: $f^{-1}(x) = \frac{1}{3}x - \frac{1}{3}$)

Find the inverse function of $f(x) = 3x + 1$ and draw the graphs of $f(x)$ and $f^{-1}(x)$.

EXAM PRACTICE QUESTION 2.121

(ANSWER: $f^{-1}(x) = \frac{1}{2}x - \frac{5}{2}$)

Let f be such that x is mapped onto $2x+5$. Calculate the inverse function $f^{-1}(x)$ and sketch both functions.

EXAM PRACTICE QUESTION 2.122

(ANSWERS: (a) $f^{-1}(x) = \frac{x-5}{3}$, (b) $g^{-1}(x) = -x+7$, (c) $h^{-1}(x) = \frac{x+2}{5}$)

If (a) $f(x) = 3x+5$, find $f^{-1}(x)$, (b) If $g(x) = -x+7$, find $g^{-1}(x)$, (c) If $h(x) = 5x-2$, find $h^{-1}(x)$. Sketch all six graphs.

EXAM PRACTICE QUESTION 2.123 (ANSWER: $f(g(x)) = \frac{1}{x+5}$, $g(f(x)) = \frac{1}{x} + 5$)

If $f(x) = \frac{1}{x}$ and $g(x) = x + 5$, find $f(g(x))$ and $g(f(x))$.

EXAM PRACTICE QUESTION 2.124 (ANSWERS: (i) $f \circ g(x) = f(g(x)) = 3x + 4$, (ii) $g \circ f(x) = g(f(x)) = 3(x+4)$)

If $f(x) = x + 4$ and $g(x) = 3x$, Find: (i) $f \circ g\,(x) = f(g(x))$ (ii) $g \circ f\,(x) = g(f(x))$.

EXAM PRACTICE QUESTION 2.125 (ANSWERS: (i) $f(g(x)) = \frac{6}{-2x+7}$, (ii) $g(f(x)) = -\frac{12}{x} + 7$, (iii) f(g(2))=2, g(f(2))=1, g(f(3))=3)

If $f(x) = \frac{6}{x}$ and $g(x) = -\,2x + 7$,

find: (i) $f(g(x))$, (ii) $g(f(x))$ and (iii) Evaluate $f(g(2))$, $g(f(2))$ and $g(f(3))$

Topic 2.22 TRIAL AND IMPROVEMENT.

EXAM PRACTICE QUESTION 2.126 (ANSWER: $x = 2.44$)

Find a solution to $x^3 + x = 17$ to 2 d.p.

EXAM PRACTICE QUESTION 2.127 (ANSWER: $x = 2.83$)

Find a solution to $x^3 - 2x = 17$ to 2 d.p.

EXAM PRACTICE QUESTION 2.128 (ANSWER: $x = 2.837$)

Find a solution of the cubic equation $x^3 - x - 20 = 0$ to 3 d.p.

Topic 2.23 NUMERICAL ITERATION.

EXAM PRACTICE QUESTION 2.129 **(ANSWER: $x = 4.79$)**

Rewrite the quadratic equation $x^2 - 5x + 1 = 0$ in the form $x_{n+1} = \sqrt{ax_n + b}$ stating the values of a and b.

Using $x_1 = 1$, find a root of $x^2 - 5x + 1 = 0$ correct to 2 d.p.

EXAM PRACTICE QUESTION 2.130 (ANSWER: $x_a = -4.65$ and $x_b = 0.65$)

Use an iterative method to find the solution of $x^2 + 4x - 3 = 0$.

Topic 2.24 SEQUENCES.

Arithmetic Sequences

EXAM PRACTICE QUESTION 2.131 (ANSWER: $T_n = 5n + 2$)

Find a formula, in terms of n (where n is the number of the term), for the sequence

7, 12, 17, 22, . . .

EXAM PRACTICE QUESTION 2.132 (ANSWER: $T_n = -3n + 4$)

Find a formula, in terms of n (where n is the number of the term), for the sequence

1, –2, –5, –8, . . .

EXAM PRACTICE QUESTION 2.133 (ANSWER: $T_n = 4n + 1$)

Find the nth term of the arithmetic sequence

5, 9, 13, 17, 21, . . .

EXAM PRACTICE QUESTION 2.134 (ANSWER: $T_n = -2n + 11$, the 100th term is - 189)

Find the nth term of the arithmetic sequence 9, 7, 5, 3, . . . and write the 100th term.

EXAM PRACTICE QUESTION 2.135 (ANSWER: $T_n = 4n - 1$)

Find the nth term of the sequence

3, 7, 11, 15, 19, . . .

EXAM PRACTICE QUESTION 2.136 (ANSWERS: $T_n = 2n + 1$, 3, 5, 7, 9, 11)

The first term of an arithmetic sequence is 3 and the common difference is 2.
Find the nth term and hence write down its first five terms.

EXAM PRACTICE QUESTION 2.137 (ANSWERS: $T_n = 7n - 2$, 5, 12, 19, 26, 33)

The nth term of an arithmetic sequence is $T_n = a + (n-1)d$. If the first term is 5 and the common difference 7, find the formula of its nth term and hence write down the first five terms.

Quadratic Sequences

Exam Practice Question 2.138 (Answer: The second differences are constant and equal to $2a$)

In $T_n = an^2 + bn + c$ substitute $n = 1, 2, 3, 4, 5$ and generate the first five terms of the quadratic sequence. Then write down the general sequence and look at the differences. Determine the second differences.

Exam Practice Question 2.139 (Answer: $T_n = n^2 - n + 9$)

Find the rule that generates this sequence

9, 11, 15, 21, 29, 39, ...

EXAM PRACTICE QUESTION 2.140 (ANSWER: $T_n = -n^2 + 6n + 4$)

Find the general rule for the sequence

9, 12, 13, 12, 9, 4, …

EXAM PRACTICE QUESTION 2.141 (ANSWER: $T_n = n^2 + n + 1$)

Find the nth term of the sequence

3, 7, 13, 21, 31, ...

Geometric sequence

EXAM PRACTICE QUESTION 2.142 **(ANSWERS: 2, 4, 8, 16,... $T_n = 2\times 2^{n-1}$)**

The first term of geometric sequence is 2 and the common ratio, $r = 2$.
Write down the first few terms and hence find the nth term.

EXAM PRACTICE QUESTION 2.143 **(ANSWERS: (a) r=2, (b) the 5th term is 48)**

A geometric progression is given as 3, 6, 12, 24, 48

(a) Find the common ratio, r.

(b) Deduce that the 5th term is 48, using the nth term formula ar^{n-1}.

Fibonacci (Leonardo Pisano) Sequence

This sequence is presented in the book «GCSE (9-1) Mathematics Higher»[5].

[5] Dr Gerasimos Masouros & Anthony Nicolaides, «GCSE (9-1) Mathematics Higher», PASS Publications Ltd, London 2018, ISBN 9781872684512

Topic 2.25 TRIANGULAR NUMBERS.

Exam Practice Question 2.144

Gauss was a famous Engineer and Mathematician. When he was 11 years old in Germany, the teacher asked his class to find the sum of $1+2+3+4+\cdots+100$, Gauss came up with the correct answer in seconds.

Can you imagine his method?

Gauss's method is presented in the book «GCSE (9-1) Mathematics Higher»[6].

[6] Dr Gerasimos Masouros & Anthony Nicolaides, «GCSE (9-1) Mathematics Higher», PASS Publications Ltd, London 2018, ISBN 9781872684512

3 RATIO, UNITS, PROPORTIONALITIES, PERCENTAGES & RATES OF CHANGE

The difficulty level of the following exam-style questions and problems is progressively increasing and their full, step-by-step solutions with the required theoretical notes, diagrams and figures can be found in the book «GCSE (9-1) Mathematics Higher»[1], pp. 167-218. The numbering of the questions has remained the same, so, for example, the full solution of «Exam Practice Question 3.1» can be found in «Worked Example 3.1.» of the above book. With this arrangement, it is simple for the students to read, understand and remember the theory and also to check and correct their work.

Topic 3.1 RATIO.

EXAM PRACTICE QUESTION 3.1 **(ANSWERS: (i) 1:4, (ii) 2:3, (iii) 5:9, (iv) 5:2, (v) 7:3, (vi) 9:5)**

Write down the corresponding ratios to the following fractions:

(i) $\frac{1}{4}$ (ii) $\frac{2}{3}$ (iii) $\frac{5}{9}$ (iv) $\frac{5}{2}$ (v) $\frac{7}{3}$ (vi) $\frac{9}{5}$

EXAM PRACTICE QUESTION 3.2 **(ANSWERS: (a) 1:2, (b) 1:4, (c) 1:5, (d) 1:5, (e) $1:\frac{1}{6}$, (f) $1:\frac{1}{7}$, (g) $1:\frac{1}{10}$, (h) $1:\frac{3}{2}$, (i) $1:\frac{6}{7}$)**

Simplify the following ratios and write them in the form $1:n$, where n is a rational number.

(a) $3:6$ (b) $2:8$ (c) $5:25$ (d) $7:35$ (e) $36:6$ (f) $21:3$ (g) $5:\frac{1}{2}$ (h) $2:3$ (i) $7:6$

[1] Dr Gerasimos Masouros & Anthony Nicolaides, «GCSE (9-1) Mathematics Higher», PASS Publications Ltd, London 2018, ISBN 9781872684512

EXAM PRACTICE QUESTION 3.3 **(ANSWERS: (a) 1:0.4, (b) 1:6, (c) 1:0.25, (d) 1:4, (e) 1:2, (f) 1:0.02)**

Write the following ratios in the form $1:n$

(a) 5 hours : 2 hours
(b) 30 minutes : 3 hours
(c) £2 : 50 p
(d) 75 p : £3
(e) 5 m : 10 m
(f) 1 kg : 20 g

EXAM PRACTICE QUESTION 3.4 (ANSWERS: (a) 3:1, (b) 4:1, (c) 1:5, (d) $1:\frac{2}{5}$, (e) $1:\frac{3}{2}$, (f) $\frac{13}{7}:1$)

Convert the following ratios into the form $1:n$ or $n:1$

(a) 900 volts : 300 volts
(b) 20 amperes : 5 amperes
(c) 1 000 turns : 5 000 turns
(d) $5:2$
(e) $4:6$
(f) $13:7$

Ratios used for scaling up or down

EXAM PRACTICE QUESTION 3.5 (ANSWER: 1:500)

The height of the Eiffel tower in Paris is 105 m, it is required to make a scale model of 21 cm. Find the ratio of the height of the model to the actual height. Give the answer in the form $1:n$

EXAM PRACTICE QUESTION 3.6 (ANSWERS: (a) 4.75 m, (b) 2.35 cm)

A model of a car is made using a scale of 1:100 .

(a) The length of the model is 4.75 cm, work out the length of the real car, give your answer in metres.

(b) The width of the real car is 2.35 m, work out the width of the model.

EXAM PRACTICE QUESTION 3.7 (ANSWER: 5 cm)

The diagram of Fig. 3.1.1 shows the distance between two cities, London and Birmingham.

The actual distance is 100 km and the scale represents 1 cm = 20 km, find the value of x.

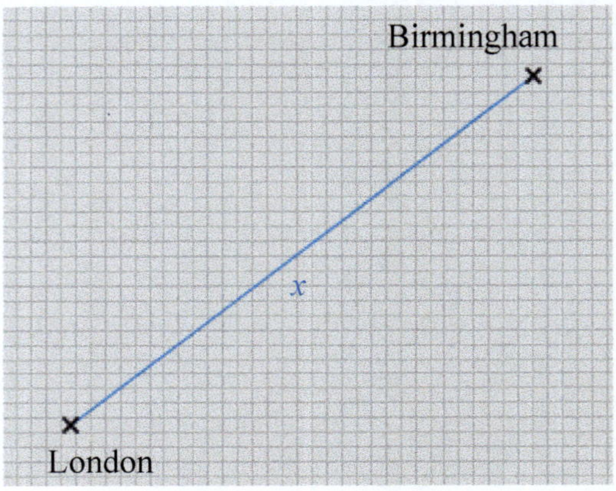

FIG. 3.1.1 (not to scale)

EXAM PRACTICE QUESTION 3.8 (ANSWERS: (a) 350 km, (b) $AC = 320$ km, $BC = 140$ km)

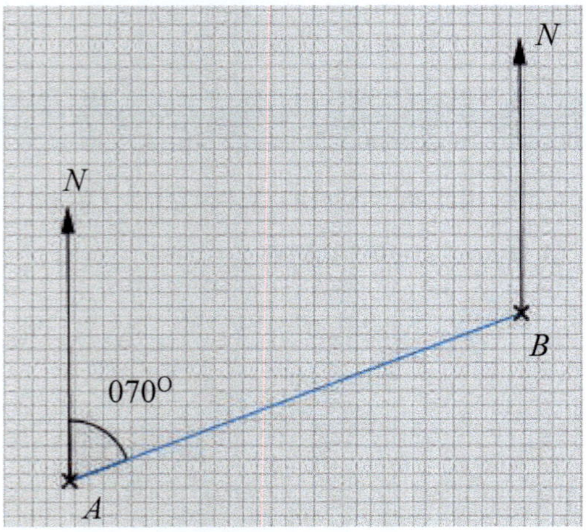

FIG. 3.1.2 (not to scale)

The diagram of Fig. 3.1.2 shows the relative positions of two airports A and B
1 cm represents 50 km, that is, the scale is :
1 cm : 50 km

(a) Given that distance AB is 7 cm on the diagram, how far away from airport A is airport B?

(b) The bearing of an airport C from A (*i.e.*, the clockwise angle between AC and the North cardinal) is 045° and the bearing for an airport C from B (*i.e.*, the clockwise angle between BC and the North cardinal – see Topic 4.18 in Geometry) is 315°. How far away from airport A and B is airport C? Give your answer in kilometres.

EXAM PRACTICE QUESTION 3.9 (ANSWER: 40 cm)

A map is drawn using a scale of $1:500\ 000$. The distance between two cities is 200 km. Calculate the distance between the two cities in cm, on the map.

EXAM PRACTICE QUESTION 3.10 (ANSWERS: (a) 1:2:3, (b) 1:2:3, (c) 4:3:2:1, (d) 10:2:1, (e) 4:3:1, (f) 2:1:2)

Simplify the following ratios:

(a) $2:4:6$ (b) $10:20:30$ (c) $12:9:6:3$

(d) $80:16:8$ (e) $\sqrt{64}:\sqrt{36}:2$ (f) $12:6:12$

EXAM PRACTICE QUESTION 3.11 (ANSWERS: (a) 9:7, (b) $1:\frac{7}{9}$)

In a year 10 classroom there are 14 girls and 18 boys. (a) Write down the fraction of the boys to the girls. Give your ratio in its simplest form. (b) Write your answer in question (a) in the form $1:n$.

Ratios used in emphasising a part of the whole, or sharing a quantity in a given way.

EXAM PRACTICE QUESTION 3.12 (ANSWERS: (a) £10, £15, (b) 30 m, 70 m, (c) 21 kg, 84 km)

Divide the quantities in the ratios given:

(a) £25 in the ratio $2:3$ (b) 100 m in the ratio $3:7$ (c) 105 kg in the ratio $1:4$

EXAM PRACTICE QUESTION 3.13 (ANSWER: £120, £40)

Jack and Gill share £160 in the ratio $3:1$. How much each receives?

EXAM PRACTICE QUESTION 3.14 (ANSWERS: 20 boys, 10 girls)

There are 30 students in a mixed class. The ratio of the boys to the girls is $2:1$. How many girls are there? What is the ratio of the girls to the total number?

EXAM PRACTICE QUESTION 3.15 **(ANSWERS: (a) 45 boys, (b) 30 girls)**

(a) The ratio of girls to boys at a party is $2:3$. There are 30 girls at the party. How many boys are at the party?

(b) The ratio of boys to girls at another party is $7:6$. There are 35 boys at the party. How many girls are at the party?

EXAM PRACTICE QUESTION 3.16 **(ANSWERS: (a) 18 boys and 12girls, (b) 16 boys, (c) 3:4)**

(a) In a mixed class of Mathematics there are 30 pupils. The ratio of boys to girls is $3:2$ How many boys and how many girls are there?

(b) In another mixed class of Mathematics there are 28 pupils. The ratio of girls to boys is $3:4$. How many boys are there?

(c) Find the ratio of the girls in class (a) to boys in class (b) in its simplest form.

Share a quantity in more than two parts, in a given ratio

EXAM PRACTICE QUESTION 3.17 **(ANSWER: £200, £400, £400)**

Share £1 000 between three people in the ratio $1:2:2$

EXAM PRACTICE QUESTION 3.18 **(ANSWER: $A1 = 263.19$ km, $A2 = 175.46$ km, $A3 = 438.65$ km)**

The distance from Aberdeen to London is 877.3 km, three adults $A1$, $A2$ and $A3$ share the driving in the ratio $3:2:5$

Work out how many km is each share, or, equivalently, how many km each drive.

EXAM PRACTICE QUESTION 3.19 **(ANSWERS: (a) 100€, (b) 30€)**

Alberto, Beryl and Clive are going on holidays to Europe by car and they will share the cost of the petrol in the ratio $2:3:5$. Alberto and Clive together pay €70.

(a) How much was the petrol?

(b) What fraction of the cost was Beryl's?

EXAM PRACTICE QUESTION 3.20 **(ANSWER: son £1 000, mother £2 000, father £3000)**

The father in a family of three (Father, Mother, Son) shared £6 000 in the ratio $3:2:1$, the smallest amount was given to the son and he kept the biggest for himself. How much did each receive?

Topic 3.2 STANDARD UNITS AND COMPOUND UNITS. CONVERSIONS BETWEEN RELATED COMPOUND UNITS

EXAM PRACTICE QUESTION 3.21 (ANSWER: 19.3×10^3 kg)

Given that each cm^3 of gold weighs 19.3 g, how much would the 1 m^3 of gold weigh?

EXAM PRACTICE QUESTION 3.22 (ANSWER: 0.04 N)

If a pressure of 20 N/m^2 is applied on a surface area of 20 cm^2 as shown in the following figure, calculate the force applied.

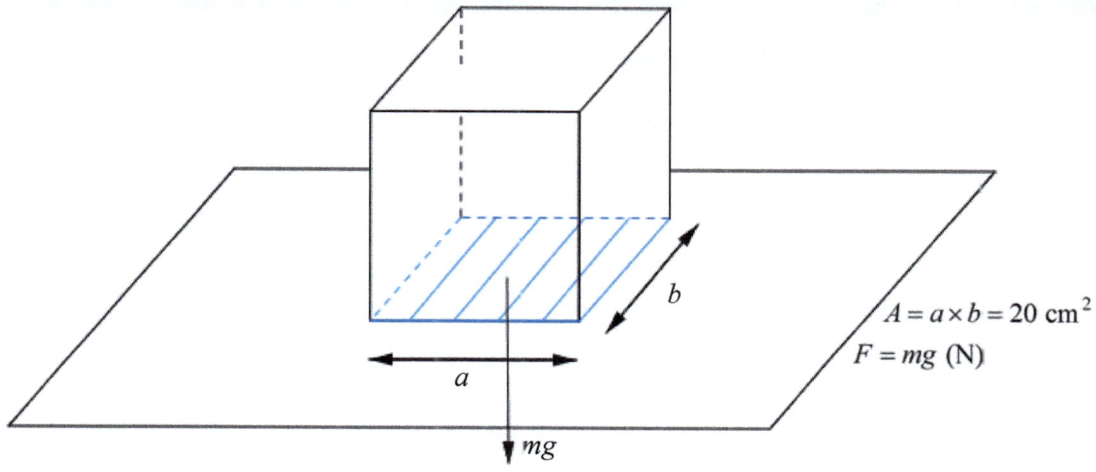

EXAM PRACTICE QUESTION 3.23 (ANSWERS: (a) $\frac{\pi}{4}\times 10^{-6}$ m^2, (b) 1.27×10^4 $\frac{\text{N}}{\text{m}^2}$)

(a) Work out the area of a circle of diameter 1 mm, in m^2.

(b) Given that a force 0.01 N is applied on the above circular area, work out the pressure on the circle to 3 significant figures.

EXAM PRACTICE QUESTION 3.24 (ANSWER: 9.81 N)

Work out the force applied on the scales when you put mass equal to 1 kg on it. Take $g = 9.81$ m/s^2

Topic 3.3 DIRECT PROPORTIONALITIES.

EXAM PRACTICE QUESTION 3.25 (ANSWER: YES, because $y = 5x$)

The table of values for the variables x and y is given below:

x	1	3	5	7	9	11
y	5	15	25	35	45	55

(a) Plot the graph of y against x.

(b) Is y directly proportional to x? Give a reason for your answer.

EXAM PRACTICE QUESTION 3.26 (ANSWERS: (a) $V = R \times I$, (b) $R = 5\,\Omega$, (c) $I = 10$ mA, (d) $R = 10\,\text{k}\Omega$)

Ohm's law states that the potential difference across a linear resistor is directly proportional to the current flowing through the resistor, provided the temperature is kept constant.

(a) Write the above statement in the proportionality form and hence give the equation, taking V as the potential difference (voltage), I as the current and the constant k as the resistance, R,

(b) If, for some resistance R, $I = 2$ A and $V = 10$ V, find R.

(c) If for another resistance $R = 1\ 500\ \Omega$, $V = 15$ volts, find the current I.

(d) For a third resistance R, draw the graph of V against I for the following sets of values of currents; then determine R.

V (volt)	10	20	30	40	50
I (mA)	1	2	3	4	5

EXAM PRACTICE QUESTION 3.27

(ANSWER: The graph of F against x is a perfect straight line with gradient k = 5 kg/mm)

Hooke's law states that the extension of a uniform wire (an elastic material) is directly proportional to the applied load, provided the limit of proportionality has not been exceeded. The following table shows the experimental data we gathered for a particular wire:

x: (extension in mm)	1	2	3	4	5
F: (load in kg)	5	10	15	20	25

It is required to plot the load F along the y-axis and the extension x along the x-axis. What is your conclusion?

EXAM PRACTICE QUESTION 3.28 (ANSWER: $R = 5\ \text{k}\Omega$)

The current through a resistor is measured by an ammeter placed in series with the resistor and the voltage or potential difference across the resistor is measured by a voltmeter placed in parallel with the resistor. In an experiment, the following sets of readings were recorded for different voltage adjustments:

I (μA)	20	40	50	60	80	90	100	120
V (mV)	100	200	250	300	400	450	500	600

Find the value of the resistance of the resistor and plot V against I.

EXAM PRACTICE QUESTION 3.29 (ANSWER: $x = \pm\ 6.3$)

If y is directly proportional to the square of x and given that $y = 10$ when $x = 2$, find x when $y = 100$, giving your answer to 2 s.f.

EXAM PRACTICE QUESTION 3.30 (ANSWER: $y = 3x^3$, $y = 192$)

If y is directly proportional to the cube of x and given that $y = 81$ when $x = 3$, find the formula of y in terms of x. If $x = 4$, find y.

EXAM PRACTICE QUESTION 3.31 (ANSWER: $y = 14$)

If y is directly proportional to the square root of x and given that $y = 12$ when $x = 36$, find the formula of y in terms of x. If $x = 49$, find y.

Topic 3.4 INDIRECTLY PROPORTIONAL OR INVERSELY PROPORTIONAL OR INDIRECT PROPORTIONALITIES.

EXAM PRACTICE QUESTION 3.32 **(ANSWER: $y = 160$)**

If y is inversely proportional to the cube of x and given that $y = 20$ when $x = 2$, find y when $x = 1$.

EXAM PRACTICE QUESTION 3.33 **(ANSWERS: (a) 48 hours, (b) 64 hours)**

It took 6 brick layers to build a wall in 32 hours.
Work out how long it would have taken to build this wall with (a) 4 brick layers, (b) 3 brick layers.

EXAM PRACTICE QUESTION 3.34 **(ANSWER: 6 hours)**

A school is cleaned in 4 hours by 6 cleaners. How long will it take 4 cleaners to clean the school.

Topic 3.5 THE METHOD OF SOLVING FOR AN UNKNOWN QUANTITY USING PROPORTIONS OF THREE KNOWN QUANTITIES.

The method for directly proportional quantities

EXAM PRACTICE QUESTION 3.35 **(ANSWER: £9.95)**

15 copies of a certain book cost £149.25. Find the cost of one copy.

EXAM PRACTICE QUESTION 3.36 **(ANSWER: 63 bottles)**

25 bottles hold 20 litres of wine. How many bottles of the same size will hold 50.4 litres wine?

The method for indirectly proportional quantities

EXAM PRACTICE QUESTION 3.37 (ANSWER: $2.6\dot{3}\ \frac{m}{s^2}$)

A man of mass 90 kg is driving his car of mass 700 kg and accelerates to the maximum ability of its engine at $3\ m/s^2$. Next day, he is driving his car in the same road with his wife and children. The mass of all the passengers is 200 kg. If the engine produces exactly the same force, what is the new acceleration of the car?

EXAM PRACTICE QUESTION 3.38 (ANSWERS: (i) 0.1 kg, (ii) 7.35 $m\ell$)

A glass is filled up with $100\ m\ell$ of distilled water of density 1 000 kg/m^3.

(i) How much is the mass of the water in the glass?

(ii) How much volume of mercury, which has density equal to 13.6 gr/cm^3 would have exactly the same mass as the distilled water above?

Topic 3.6 PERCENTAGES (%).

(i) Find the percentage of a part of a quantity

EXAM PRACTICE QUESTION 3.39 **(ANSWER: 24 m)**

Find 20% of 120 m.

EXAM PRACTICE QUESTION 3.40 **(ANSWER: 160 kg)**

Find $66\frac{2}{3}\%$ of 240 kg.

EXAM PRACTICE QUESTION 3. 41 **(ANSWERS: (i) 86.25, (ii) 76.95)**

(i) Increase 75 by 15%, (ii) Decrease 95 by 19%

(ii) Write a quantity (A) as a Percentage (%) of another quantity (B)

EXAM PRACTICE QUESTION 3.42 **(ANSWER: 40%)**

Express $\frac{2}{5} = 2 \div 5$ as a percentage.

Exam Practice Question 3.43 (Answer: 37.5%)

Write 75 p as a percentage of £2.

Exam Practice Question 3.44 (Answer: 500%)

Find 125 as a percentage of 25.

Exam Practice Question 3.45 (Answers: (i) 150%, (ii) 50%)

A load of 750 kg is allowed to be carried by a car. Traffic Control Police found the driver to carry 1 125 kg with his car. (i) Write 1 125 kg as a percentage of 750 kg. (ii) By what percent did the driver exceed the limit?

Exam Practice Question 3.46 (Answers: (i) 300%, (ii) 16%, 2%)

Alice who is 60 kg is driving her car, which is 820 kg to the supermarket, where she picks her husband Anthony, who is 100 kg and his shopping bags, which are 20 kg altogether.

(i) Calculate the percentage increase of the weight that the car is carrying, after Anthony entered with his shopping in Alice's car.

(ii) What percentage of the total weight is the weight of the passengers and what percentage of the total weight is the weight of Anthony's shopping bags?

Exam Practice Question 3.47 (Answer: £45)

The cost of a dress is £60 before the sale discount of 25% is removed. How much do you pay?

EXAM PRACTICE QUESTION 3.48 (ANSWER: A = £68, B = £48)

The Original price of a pair of trousers, A and that of a pair of shoes, B had a ratio 4 : 3. During a sale, A is reduced by 15% and B is reduced by 20%. If the original price of A is £80, what is the sale price of A and B?

EXAM PRACTICE QUESTION 3.49 (ANSWER: £141, or 23.5%)

There is a sale for men's suits of 15% and the last day of the sale there was a further 10% discount on the already discounted prices. Find the total percentage discount on the original suit price. If the original price was £600, how much is the total discount?

EXAM PRACTICE QUESTION 3.50 (ANSWER: original price = £2 000, amount paid = £1 400)

The price of a computer in a sale was reduced by £600. The reduction was 30%. Find the original price and the amount paid.

EXAM PRACTICE QUESTION 3.51 (ANSWER: ℓ = 25 cm)

A uniform steel cuboid has length ℓ cm, width 10 cm and height 5 cm. The ratio of the pressure exerted on the ground when stood on face A, to when it stood on face B (see figure below) is 5 : 2.
Find the length ℓ.

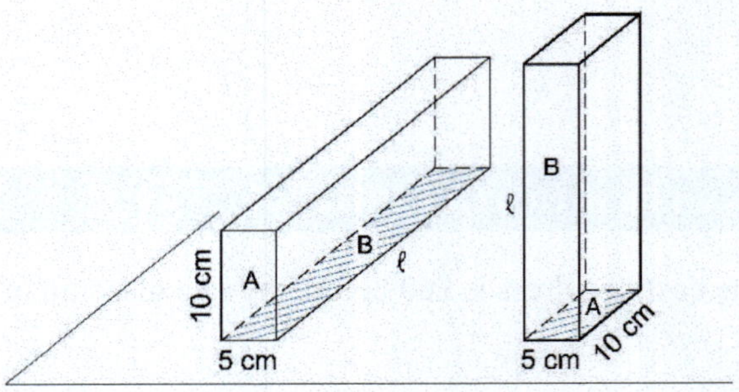

FIG. 3.6.1

(iii) Finding the Net and Gross Amounts

EXAM PRACTICE QUESTION 3.52 (ANSWER: £76.20)

The net cost of a camera is £63.50. If V.A.T. of 20% is applied, what is its gross cost?

EXAM PRACTICE QUESTION 3.53 (ANSWERS: (i) £81.35, (ii) 6.76%)

The camera of the previous worked example costs £63.50 plus V.A.T of 20% or you could pay a 10% of the net cost deposit followed by 6 monthly payment of £12.50. (i) How much is the gross cost with the monthly payments method? (ii) By what percentage it has changed, compared to the payment in cash method?

EXAM PRACTICE QUESTION 3.54 (ANSWERS: (i) £420, (ii) £350)

Gross cost of a dress is £357 in the sale after a reduction of 15%. (i) Find the original gross cost and (ii) calculate the net cost of the dress, considering VAT of 20%.

EXAM PRACTICE QUESTION 3.55 **(ANSWER: food = £54, wine = £22.50)**

Anthony bought food and wine in the supermarket and paid £81. Considering that the wine has 20% VAT, while food doesn't have VAT and that the rate of the gross cost of the wine to the gross cost of the food was 1 : 2 what is the net cost of each one (food and wine)?
Consider no overheads other than the VAT

(iv) Simple Interest

EXAM PRACTICE QUESTION 3.56 **(ANSWER: £300)**

Calculate the simple interest earned on £2 000 at 5% p.a. for 3 years.

(v) Compound Interest

EXAM PRACTICE QUESTION 3.57 **(ANSWER: £632.22)**

£450 is invested at 12% p.a., compound interest, for 3 years. How much is the total amount at the end of year 3?

EXAM PRACTICE QUESTION 3.58 **(ANSWER: £5 329.95)**

A car depreciates at 11% of its current prize p.a. Presuming that it was bought for £8 495, what is its value after 4 years?

(vi) Percentage Increase (↑) or Decrease (↓)

EXAM PRACTICE QUESTION 3.59 **(ANSWER: 13.6%)**

A house cost £110 000 in 1999 and £125 000 in 2000. What is the percentage increase of its value?

EXAM PRACTICE QUESTION 3.60 **(ANSWERS: (i) $25.70\times10^3\ \frac{N}{m^2}$, (ii) $22.56\times10^3\ \frac{N}{m^2}$, (iii) $90.24\times10^3\ \frac{N}{m^2}$, (iv) $12.83\times10^3\ \frac{N}{m^2}$, (v) $P_{ii}\% = 87.78\%$, $P_{iii}\% = 351.13\%$, $P_{iv}\% = 49.92\%$)**

A uniform steel cone with density $\rho = 7.70 \times 10^3$ kg/m^3, height 1 m and radius of the base 0.2 m is cut as shown in Fig. 3.6.2, so that height of the new cone : height of the original cone = 1 : 2.

(i) Calculate the pressure applied on the plane from the original cone

(ii) Calculate the new pressure applied from the frustum cone, marked as cone A in Fig. 3.6.3. (after we remove cone B from the top.

(iii) Calculate the pressure applied from cone C on the plane, that is cone A turned upside down.

(iv) Calculate the pressure applied from cone B on the plane

(v) Calculate the percentage change of the pressures in (ii) to (iv) compared to initial pressure applied on the plane form the original cone. Comment on the results.

Assume that $g = 10$ m/s^2.

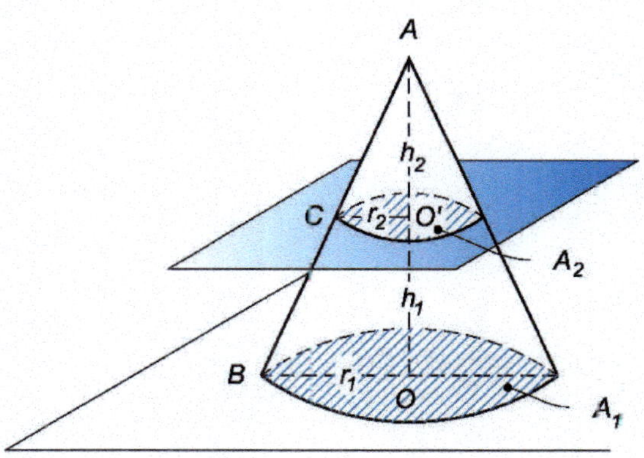

FIG. 3.6.2.

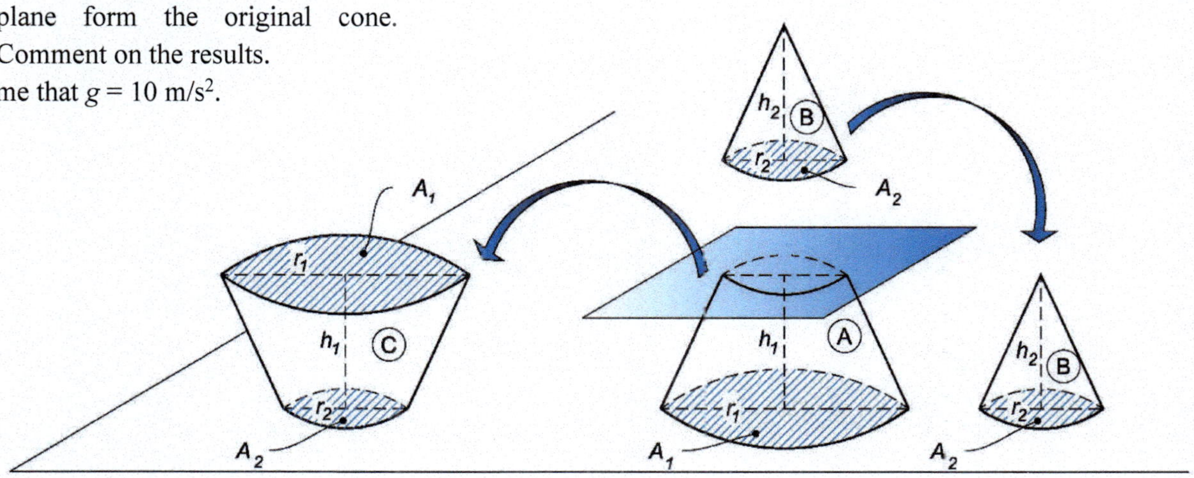

FIG. 3.6.3.

Topic 3.7 RATES OF CHANGE.

EXAM PRACTICE QUESTION 3.61 **(ANSWER: the pole vault landing system decreases significantly the deceleration and consequently the generated force on the human body)**

In the history of pole vault, Sergey Bubka was the first athlete to clear 6.00 metres and 6.10 metres and kept raising his record for many years, breaking the world record for men's pole vault 35 times and beating his own record 14 times, as he beat his record for 14 consecutive times, without any other athlete interrupting this process.

6.00 metres is as high as the second floor of a building. Explain why a pole vaulter of mass $m = 100$ kg is not hurt when he is falling from a height of 6 metres, while any man falling form a second-floor balcony hurts himself severely or even dies.

EXAM PRACTICE QUESTION 3.62 **(ANSWERS: (a) \$14.93/annum, (b) – \$85.65/annum, (c) – \$1 265.53/annum, (d)) \$386.28/annum)**

The UK GDP (that is Gross Domestic Product) per capita, in USD, that is the average income per person, for the period 2006 to 2014 is shown in the following bar chart

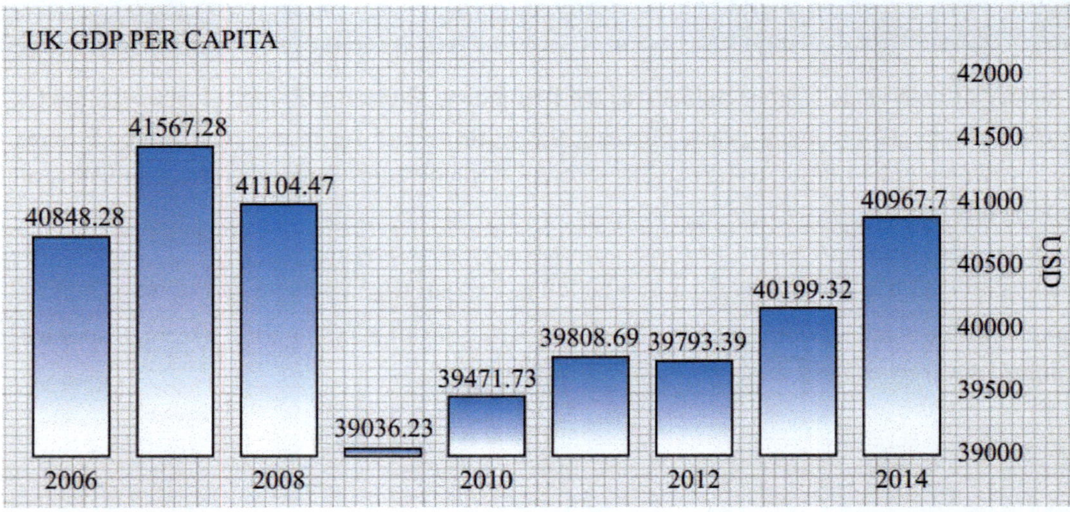

FIG. 3.7.1

Determine the rate of change:

(a) between the years 2006 and 2014

(b) between the years 2007 and 2014

(c) between the years 2007 and 2009

(d) between the years 2009 and 2014

EXAM PRACTICE QUESTION 3.63 **(ANSWER: rate of change of A is 2 and of B is 7)**

Assume that the functions $y = 2t$ and $y = t^2$ are describing the rate of change of quantities A and B respectively. Compare the two rates of change and determine which one is bigger.

Hint: Construct one table of values for both of them and compare their rates of change. Alternatively draw their graphs on the same figure.

EXAM PRACTICE QUESTION 3.64 **(ANSWER: (a) 2, (b) 2, they are the same, because they both represent the slope of the same straight line)**

Draw the graph of the function $y = 2t + 5$ for $t = 0$ s to $t = 10$ s and find the rate of change

(a) between $t = 1$ s and $t = 2$ s

(b) between $t = 0$ s and $t = 10$ s

Give reasons

EXAM PRACTICE QUESTION 3.65 (ANSWERS: (b) 1.1 m/s, (c) 1 s)

A particle is moving with acceleration of $1\ \mathrm{m/s^2}$ and at $t_0 = 0\ \mathrm{s}$ its speed is $0.1\ \mathrm{m/s}$

a) Draw the graph of distance over time for the first two seconds (for time $t = 0$ to $t = 2$).

b) Work out the average rate of change of the distance (= average speed) of the particle from the origin (= its position at $t_0 = 0\ \mathrm{s}$) during the first 2 seconds.

c) From the graph determine the time t when the instantaneous rate of change of its distance from the origin (= instantaneous speed) is equal to the average distance (speed).

EXAM PRACTICE QUESTION 3.66 (ANSWERS: $-0.172\ \frac{\text{mm}}{\text{year}}$, $-0.467\ \frac{\text{mm}}{\text{year}}$)

When an iron pipe is buried in the soil, unprotected, it starts rusting and the decrease of its wall thickness is exponential. A 10 mm wall thickness pipe was buried in soil and put in operation while a monitoring system recorded the following data:

Pipe wall thickness (mm)	10	9.89	9.78	9.65	9.51	9.35	9.18	8 99
Time (years)	0	1	2	3	4	5	6	7

Continued below

8.77	8.54	8.28	8.00	7.68	7.33	6.94	6.52	6.05
8	9	10	11	12	13	14	15	16

Continued below

5.53	4.95	4.31	3.61	2.83	1.97	1.03
17	18	19	20	21	22	23

Data was extrapolated (that is estimated beyond the monitored period) up to 30 years, as follows:

Pipe wall thickness (mm)	– 0.02	– 1.18	– 2.46	– 3.88	– 5.44	– 7.17	– 9.09
Time (years)	24	25	26	27	28	29	30

TABLE 3.7.1

a) Draw the graph of pipe wall thickness in mm over time (in years) for $t = 0$ to $t = 30$ years.

b) Work out the average rate of change of the pipe wall thickness the first 10 years and the average rate of change during the period 10 to 20 years. Comment on your results.

EXAM PRACTICE QUESTION 3.67 **(ANSWERS: (b) $9.8 \approx 10 \frac{\text{rabbits}}{\text{month}}$, $0.5 \approx 0 \frac{\text{rabbits}}{\text{month}}$, (c) 3.8 months and 7.9 months, (d) during the period 0 to 2 months (beginning of the experiment) the rate of change equals to the rate of change during the period 10 to 12 months, (e) 1%, 0.99%, 5.88%, 11.11%, 16.67%, 17.14%, 10.98%, 5.49%, 2.60%, 0.51%, 0.51%, 0%**

In nature, the course of a self-limiting processes such as the growth of an organism or population in a limited environment are best described by the exponential functions, like the one presented in Chapter 2, Topic 2.17, of the book «GCSE (9-1) Mathematics Higher»[2], that is functions which change rapidly.
For example, if two rabbits were placed on an island with no predators but a limited food supply, the number of rabbits would increase exponentially for a time (for as long as there was plenty of food), then slow down and finally round off at the carrying capacity of the island, that is the number of rabbits the island could accommodate.

In an experiment 100 rabbits were placed in a room with limited food supply and the number of rabbits was measured once every month for 12 months. The results are presented in the following table of values:

Number of rabbits	100	101	102	108	120	140	164	182	192	197	198	199	199
Time (months)	0	1	2	3	4	5	6	7	8	9	10	11	12

TABLE 3.7.2

a) Draw the graph of the number of rabbits over time (in months) for t = 0 to t = 12 months.
b) Work out the average rate of change of the number of rabbits the first 10 months and the average rate of change during the period 10 to 12 months. Comment on your results.
c) From the graph determine the time t when the rate of change of the number of the rabbits (= instantaneous rate of change) is equal to the average rate of change for the period 0 to 10 months.
d) Repeat the above (c) for the average rate of change for the period 10 to 12 months.
e) Calculate the percentage of the growth of the population and draw the corresponding bar chart. Comment on it.

[2] Dr Gerasimos Masouros & Anthony Nicolaides, «GCSE (9-1) Mathematics Higher», PASS Publications Ltd, London 2018, ISBN 9781872684512

Topic 3.8 THE GOLDEN RATIO

Divide a straight-line segment in two parts a and b, such that, $a > b$ and

$$\frac{a+b \text{ (total length)}}{a \text{ (long part)}} = \frac{a \text{ (long part)}}{b \text{ (short part)}}$$

Calculate the ratio $\varphi = \dfrac{a}{b}$

ANSWER: $\varphi = \dfrac{1+\sqrt{5}}{2} = 1.61803\ldots$

4 GEOMETRY AND MEASURES

The difficulty level of the following exam-style questions and problems is progressively increasing and their full, step-by-step solutions with the required theoretical notes, diagrams and figures can be found in the book «GCSE (9-1) Mathematics Higher»[1], pp. 219-380. The numbering of the questions has remained the same, so, for example, the full solution of «Exam Practice Question 4.1» can be found in «Worked Example 4.1.» of the above book.

With this arrangement, it is simple for the students to read, understand and remember the theory and also to check and correct their work. For their convenience, we provide the numeric answers, so the students can check and correct their answers immediately. This is not possible when figures and diagrams are required. In such cases the students must refer to the corresponding Worked examples in «GCSE (9-1) Mathematics Higher»[1].

Topic 4.1 REFLECTIONS ROTATIONS TRANSLATIONS & ENLARGEMENTS OF SHAPES.

4.1.1 Reflection

EXAM PRACTICE QUESTION 4.1

Sketch the line $f(x) = x + 2$. Hence sketch $-f(x) = -x - 2$.

[1] Dr Gerasimos Masouros & Anthony Nicolaides, «GCSE (9-1) Mathematics Higher», PASS Publications Ltd, London 2018, ISBN 9781872684512

EXAM PRACTICE QUESTION 4.2

(ANSWERS: (i) $A'(2,5)$, $B'(2,8)$, $C'(5,8)$, (ii) $A''(-5,-2)$, $B''(-8,-2)$, $C''(-8,-5)$)

Plot the triangle ABC with coordinates: $A(5,2)$, $B(8,2)$ and $C(8,5)$.

(i) Reflect the triangle ABC in the line $y = x$ and write down the coordinates of the reflected triangle.

(ii) Reflect the reflected triangle $A'B'C'$ in the line $y = -x$ and write down the coordinates of the new image.

EXAM PRACTICE QUESTION 4.3

(ANSWERS: (i) $y = -2$, (ii) $y = x$)

Plot the points $A(1, 1)$, $B(2, 4)$, $C(4, 4)$, $D(5, 1)$. Join up the points and label the shape (object).
Also plot the points $A'(1, -5)$, $B'(2, -8)$, $C'(4, -8)$, $D'(5, -5)$. Join up the points and label the shape (image 1).
Next, plot the points $A''(-8, 2)$, $B''(-8, 4)$, $C''(-5, 5)$, $D''(-5, 1)$. Join up the points and label the shape (image 2).

(i) What is the equation of the line which would reflect the object onto the image 1?

(ii) What is the equation of the line that would map Image 1 onto Image 2?

EXAM PRACTICE QUESTION 4.4 (ANSWER: $A'(2,3)$, $B'(5,3)$, $C'(5,0)$ $A''(-4,1)$, $B''(-7,1)$, $C''(-7,4)$)

Plot the points $A(2, 1)$, $B(5, 1)$, $C(5, 4)$. Label the points and join them up to give the triangle ABC. Reflect the triangle to the lines $y = 2$ and, $x = -1$. Label the triangles, $A'B'C'$ (Image 1) and $A''B''C''$ (Image 2) respectively, stating their coordinates.

4.1.2 Rotation

EXAM PRACTICE QUESTION 4.5 (ANSWER: $A'(-2,2)$, $B'(-2,4)$, $C'(-5,2)$ $A''(2,-2)$, $B''(2,-4)$, $C''(5,-2)$, $A'''(-2,-2)$, $B'''(-4,-2)$, $C'''(-2,-5)$)

Plot the points $A(2, 2)$, $B(4, 2)$, $C(2, 5)$. Join the points and label the triangle ABC which is the object shape. Rotate the object by 90° counterclockwise, 180° and 90° clockwise (or, equivalently 270° counterclockwise) and determine the coordinates of the vertices of all the image triangles.

EXAM PRACTICE QUESTION 4.6 (ANSWER: $A'(3,-3)$, $B'(6,-3)$, $C'(6,-7)$, $D'(3,-7)$)

Plot the vertices of the rectangle $ABCD$ with coordinates $A(3,3)$, $B(3,6)$, $C(7,6)$ and $D(7,3)$.

Rotate this rectangle about the origin by 90° clockwise and find the coordinates of the image.

EXAM PRACTICE QUESTION 4.7 (ANSWER: $R(-2,0)$, 180^0)

The vertices of trapezium $ABCD$ have coordinates $A(2,\ 2)$, $B(4,\ 4)$, $C(8,\ 4)$, $D(10,\ 2)$ and their corresponding images, when rotated are: $A'(-6,\ -2)$, $B'(-8,\ -4)$, $C'(-12,\ -4)$, $D'(-14,\ -2)$.

Find the centre of rotation and the angle of rotation.

EXAM PRACTICE QUESTION 4.8

(ANSWERS: (a) the shape I is a reflection of the shape O over the mirror line $y = x$, (b) the shape II is a 90^0 clockwise rotation of shape O about the origin $O(0,0)$)

In the adjacent figure,

(a) Describe fully the single transformation (reflection) that will map triangle O (Object) onto triangle I (Image I).

(b) Describe fully the single transformation (rotation) that will map triangle O (Object) onto triangle II (Image II)

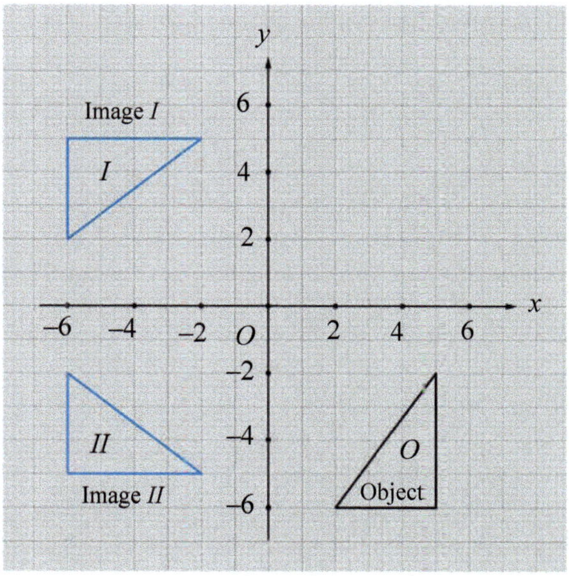

EXAM PRACTICE QUESTION 4.9

(ANSWER: $P(-2,2)$)

Plot the points $A(2, 2), B(4, 2), C(2, 6)$. Join the points and label the triangle ABC which is the object shape. Also, plot the points $A'(-2, -2), B'(-2, -4), C'(2, -2)$ and join them with straight line segments. Given that $A'B'C'$ is the shape image of the rotation of ABC around a centre, find the centre of rotation

EXAM PRACTICE QUESTION 4.10 (ANSWER: $R(0,-1)$)

The triangle ABC has coordinates $A(1, 1)$, $B(3, 1)$, $C(3, 4)$ and the corresponding images are $A'(-2, 0)$, $B'(-2, 2)$, $C'(-5, 2)$. Given that $A'B'C'$ is the image produced after a counterclockwise rotation of ABC, find the coordinates of the centre of rotation.

EXAM PRACTICE QUESTION 4.11 (ANSWER: $R(-1,-2)$, 90^0 counterclockwise)

The rectangle $ABCD$ has coordinates $A(3, 2)$, $B(3, 7)$, $C(5, 7)$, and $D(5, 2)$ and the vertices of the corresponding image are $A'(-5, 2)$, $B'(-10, 2)$, $C'(-10, 4)$, $D'(-5, 4)$.
Considering that the image is produced by an anticlockwise rotation of the object, draw the graph and find the centre of rotation and the angle of rotation.

EXAM PRACTICE QUESTION 4.12 (ANSWER: $R(\frac{1}{2},0)$, 180^0)

The vertices of the rectangle $ABCD$ (object) have coordinates $A(3, 3)$, $B(3, 6)$, $C(7, 6)$, $D(7, 3)$ and the corresponding images $A'(-2, -3)$, $B'(-2, -6)$, $C'(-6, -6)$, $D'(-6, -3)$.
Considering that the image is produced by an anticlockwise rotation of the object, draw the graph and find the centre of rotation and the angle of rotation.

EXAM PRACTICE QUESTION 4.13 (ANSWER: $R(-1,-1)$, 90^0 clockwise)

The line segment AB with coordinates $A(1, 2)$, $B(4, 4)$ is rotated and the corresponding image has coordinates $A'(2, -3)$, $B'(4, -6)$. Find the centre of rotation and the angle of rotation.

EXAM PRACTICE QUESTION 4.14 (ANSWER: $A'(3,10)$, $B'(5,14)$, $C'(1,16)$)

Plot the vertices of a triangle *ABC*, with coordinates $A(2,\ 3)$, $B(6,\ 1)$, $C(8,\ 5)$. Rotate this triangle about the centre $(-1,\ 7)$ anticlockwise by 90°. Find the coordinates of the images (graphically).

EXAM PRACTICE QUESTION 4.15 (ANSWER: $A'(0,7)$, $B'(-5,7)$, $C'(-5,9)$, $D'(0,9)$)

Plot the vertices of the rectangle *ABCD*, with coordinates $A(3,\ 2)$, $B(3,\ 7)$, $C(5,\ 7)$, $D(5,\ 2)$. Rotate this rectangle anticlockwise by 90° about the centre $R(-1,\ 3)$ and find the coordinates of the images (graphically).

4.1.3 Translation

EXAM PRACTICE QUESTION 4.16 **(ANSWER: $A'(7,4)$)**

Translate a point A (2, 2) of the first quadrant by the vector $\begin{pmatrix} 5 \\ 2 \end{pmatrix}$ and determine its new coordinates.

EXAM PRACTICE QUESTION 4.17 **(ANSWER: shape B: $(1,1.5)$, $(1,2.5)$, $(2,3)$, $(2,1)$ and shape C: $(-2,-1)$, $(-1,-1.5)$, $(-1,-2.5)$, $(-2,-3)$)**

Plot the points with coordinates (– 1, 2), (– 1, 4), (– 2, 2.5) and (– 2, 3.5) and join them to produce the trapezium A.

(a) On the grid, translate the trapezium by $\begin{pmatrix} 3 \\ -1 \end{pmatrix}$ and label the new shape by B.

(b) On the same grid, rotate trapezium B, $180°$ about $O(0,0)$ and label the new shape C.

Determine the coordinates of the shapes B and C.

EXAM PRACTICE QUESTION 4.18

(ANSWERS: (i) translation by $\begin{pmatrix}-7\\4\end{pmatrix}$, (ii) 180^0 rotation of *ABCD* about the origin)

Give the transformation that maps rectangle $ABCD$, where $A(2, 1)$, $B(6, 1)$, $C(6, 4)$, $D(2, 4)$ to:

(i) rectangle $A'B'C'D'$, where $A'(-5, 5)$ $B'(-1, 5)$, $C'(-1, 8)$, $D'(-5, 8)$

(ii) rectangle $A''B''C''D''$, where $A''(-2, -1)$, $B''(-6, -1)$, $C''(-6, -4)$, $D''(-2, -4)$.

EXAM PRACTICE QUESTION 4.19

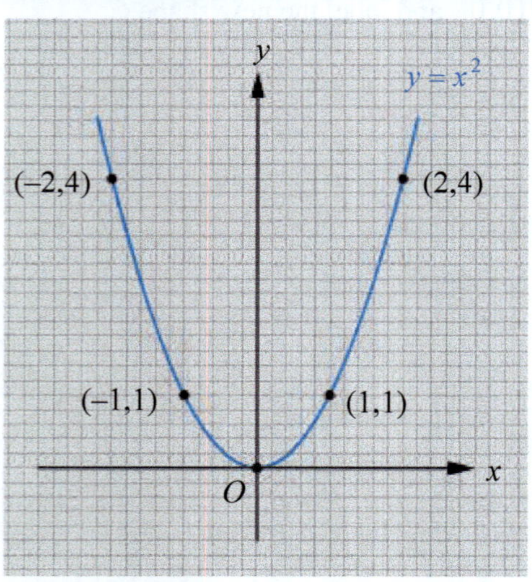

Given the adjacent graph of $y = x^2$, sketch

(i) $y = (x-3)^2$ and

(ii) $y = (x+2)^2$

EXAM PRACTICE QUESTION 4.20

Given that $y = f(x) = x^3$, sketch the curves (i) $y = x^3 + 1$ (ii) $y = x^3 - 8$.

EXAM PRACTICE QUESTION 4.21 (ANSWER: $a = -3$, $b = -1$)

Rewrite $y = x^2 - 6x + 8$ in the form $y = (x + a)^2 + b$ stating the values of a and b.

Hence starting from the curve $y = x^2$, sketch the curve $y = x^2 - 6x + 8$

4.1.4 Enlargement

(i) Enlargements by positive scale factors

EXAM PRACTICE QUESTION 4.22

Draw the object with vertices $A(3, 3)$, $B(3.5, 3)$, $C(3.5, 1.5)$, $D(4.5, 1.5)$, $E(4.5, 1)$ and $F(3, 1)$. Next, draw its enlargement with scale factor (S.F.) 2 and with centre of enlargement P which lies two units away from the object, on the x – direction, that is $P(1, 1)$.

EXAM PRACTICE QUESTION 4.23 (ANSWER: $P(-6, 3)$, S.F. = 2)

Plot the points: object $A(-4, 2)$ $B(-2, 2)$ $C(-4, -2)$

image $A'(-2, 1)$ $B'(2, 1)$ $C'(-2, -7)$.

Find the coordinates of the centre of enlargement and estimate the enlargement scale factor.

(ii) Enlargements by positive fractional scale factors (proper fraction)

EXAM PRACTICE QUESTION 4.24

Draw an enlargement of $A(-3, 4)$, $B(0, 4)$, $C(2, 1)$, D$(-4, 1)$ of S.F. $\frac{1}{3}$ and centre (6, 7).

EXAM PRACTICE QUESTION 4.25

Given the curve $y = f(x) = x^2 - x - 12$, sketch the curve $y = f(2x)$, giving coordinates where possible.

EXAM PRACTICE QUESTION 4.26

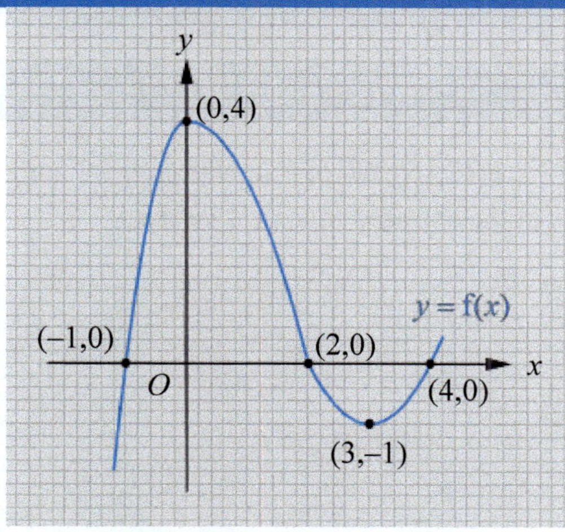

Given the adjacent figure for f(x), sketch

(i) $y = f(x - 2)$

(ii) $y = f(x) + 1$ and

(iii) $y = f(2x)$.

(iii) Enlargements by negative scale factors

EXAM PRACTICE QUESTION 4.27 (ANSWER: $A'(7,6)$, $B'(7,10)$, $C'(13,6)$)

Draw an enlargement of the triangle with vertices $A(-2, -3)$, $B(-2,-5)$, $C(-5, -3)$, with S.F. -2 and centre of enlargement $P(1, 0)$. Determine the coordinates of the enlarged triangle's vertices.

EXAM PRACTICE QUESTION 4.28 (ANSWER: $P(2,0)$, S.F. $= -\frac{1}{2}$)

Plot the points: $A(-2, 4)$, $B(-2, 12)$, $C(-6, 4)$ and $A'(4, -2)$, $B'(4, -6)$, $C'(6, -2)$.
Hence, find the centre of enlargement and estimate the S.F..

4.1.5 Combining Transformations

EXAM PRACTICE QUESTION 4.29 **(ANSWER: these translations have no invariant points)**

Draw the rectangle $ABCD$ with vertices: $A(2, 6)$, $B(4, 6)$, $C(4, 2)$, $D(2, 2)$ and label it by the word "Object"

(a) Draw the enlarged shape about the centre $P(16, 14)$ with a scale factor of $\frac{1}{2}$.

Label this shape by I.

(b) Translate the enlarged shape I by $\begin{pmatrix} -1 \\ -6 \end{pmatrix}$. Label this shape by II.

(c) Rotate II about the point (6, 2) 180°. Label this shape by the letter III.

(d) In each one of the above transformations determine the invariant points, if there exist any.

EXAM PRACTICE QUESTION 4.30 **(ANSWERS: (a) $A'(-6,3)$, $B'(-6,7)$, $C'(-4,8)$, $D'(-4,2)$**
(b) $A''(4,3)$, $B''(4,7)$, $C''(2,8)$, $D''(2,2)$,
(c) $A'''(4,-4)$, $B'''(4,0)$, $C'''(2,1)$, $D'''(2,-5)$,
(d) $A''''(5,-2)$, $B''''(5.0)$, $C''''(4,0.5)$, $D''''(4,-2.5)$)

Draw the trapezium $ABCD$ with vertices: $A(3, 6)$, $B(7, 6)$, $C(8, 4)$, $D(2, 4)$.

(a) Rotate the given trapezium $ABCD$ anticlockwise 90° about the origin. Label this new shape by I.

(b) Reflect the shape I in the line $x = -1$. Label this new shape by II.

(c) Translate the reflected shape by $\begin{pmatrix} 0 \\ -7 \end{pmatrix}$. Label this new shape by III.

(d) Enlarge the shape III about (6, 0) with a scale factor $\frac{1}{2}$. Label this new shape by IV and shade it.

Determine the coordinates of each new shape.

EXAM PRACTICE QUESTION 4.31 **(ANSWERS: (a) $A'(3,1)$, $B'(3,-3)$, $C'(5,-3)$, (b) $A''(-1,2)$, $B''(-1,-2)$, $C''(1,-2)$, (c) $A'''(-1,2)$, $B'''(-1,6)$, $C'''(1,6)$, (d) $A''' = A''$, is an invariant point in the reflection of *II* in the line $y = 2$. There exist no other invariant points in these transformations)**

Plot triangle ABC with vertices A (2, 4), B (6, 4) and C (6, 6). Then:

(a) Rotate the triangle ABC (object) clockwise 90°about the point (1, 2). Label this new shape by *I*.

(b) Translate *I* by $\begin{pmatrix} -4 \\ 1 \end{pmatrix}$. Label this new shape by *II*.

(c) Reflect *II* in the line $y = 2$. Label this new shape by *III*.

(d) On each one of the above translations, determine the invariant points, if there exist any.

Determine the coordinates of each new shape.

4.1.6 The Reciprocal Function as a Complex Transformation

EXAM PRACTICE QUESTION 4.32

Sketch the curve: $y = x^2$. Then, sketch the curve: $y = \mathrm{f}(x - 2)^2$. Finally, sketch

$y = \dfrac{1}{(x-2)^2}$ (the reciprocal curve).

EXAM PRACTICE QUESTION 4.33

Sketch the curve $f(x)=(x-1)(x-3)$ and its complex transformation $\dfrac{1}{f(x)}$ (the reciprocal curves)

Topic 4.2 ANGLES, TRIANGLES AND POLYGONS.

EXAM PRACTICE QUESTION 4.34 (ANSWER: $x = 50^0$)

Find the angle x in the adjacent figure

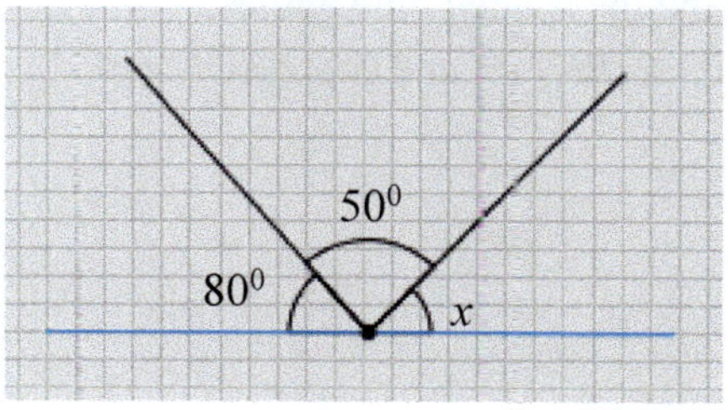

EXAM PRACTICE QUESTION 4.35 (ANSWER: $x = 60^0$)

Find the angle x in the adjacent figure.

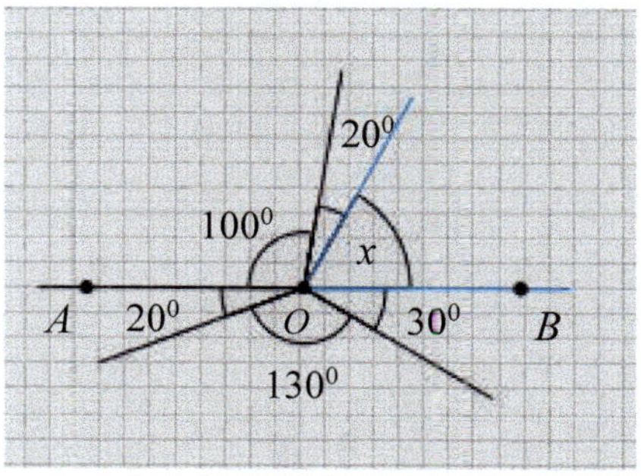

EXAM PRACTICE QUESTION 4.36 (ANSWERS: (i) $\theta = 210^0$ (ii) $\varphi = 45^0$)

In the adjacent figure, $AB = AC = BC$, $BC = CD = DE = BE$ and $ED = DF = FG$.

Also, $\measuredangle DFG = 90^0$ and the points E, D and F are on the same line.

Find the shaded blue angles:

(i) $\measuredangle ABE = \theta$ and (ii) $\measuredangle GDC = \varphi$.

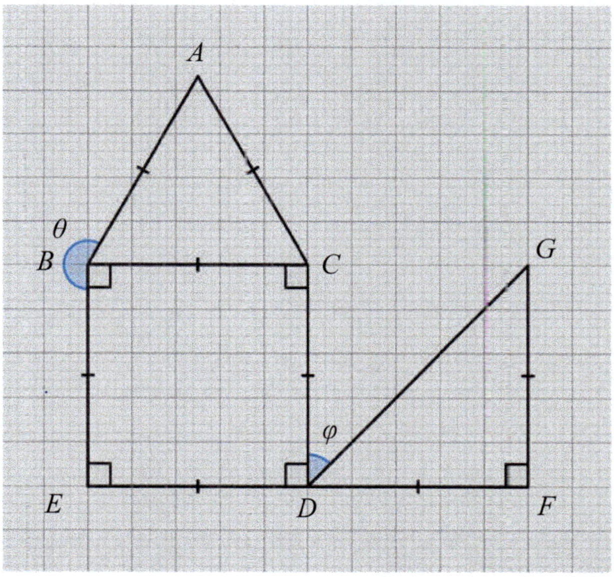

EXAM PRACTICE QUESTION 4.37 (ANSWERS: (i) $\theta = 30^0$ (ii) $\varphi = 120^0$)

In the adjacent figure $AB = AC = BC$, $BC = CD = DE = EB$ and the points B, C and F are on the same line.
Find the shaded blue angles:
(i) $\measuredangle ABE = \theta$ and (ii) $\measuredangle ACF = \varphi$.

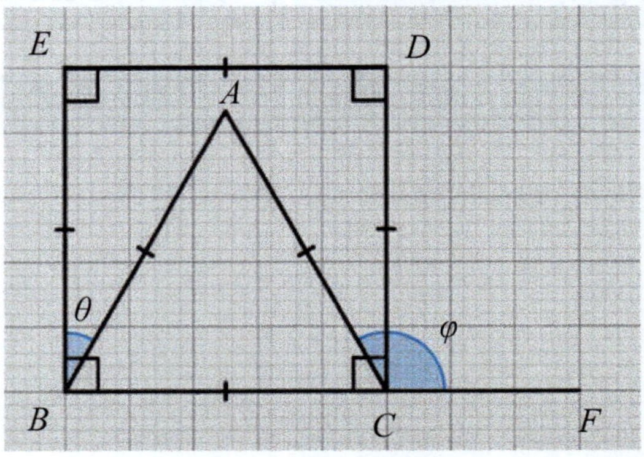

EXAM PRACTICE QUESTION 4.38 (ANSWER: $\varphi = 40^0$)

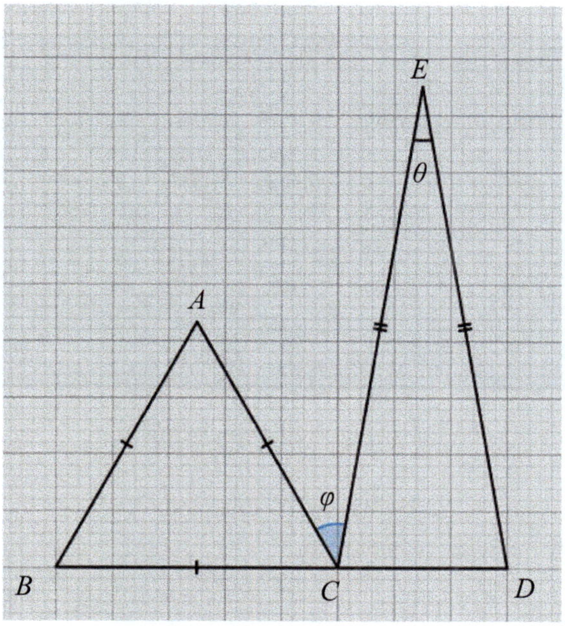

An equilateral triangle *ABC* and an isosceles triangle *ECD* share the same vertex *C*. Also, the vertices *B*, *C* and *D* are on the same line and the angle $\measuredangle CED = \theta$ is 20^0.

Find the shaded blue angle φ.

EXAM PRACTICE QUESTION 4.39 (ANSWER: $z = 20^0$, $\omega = 50^0$)

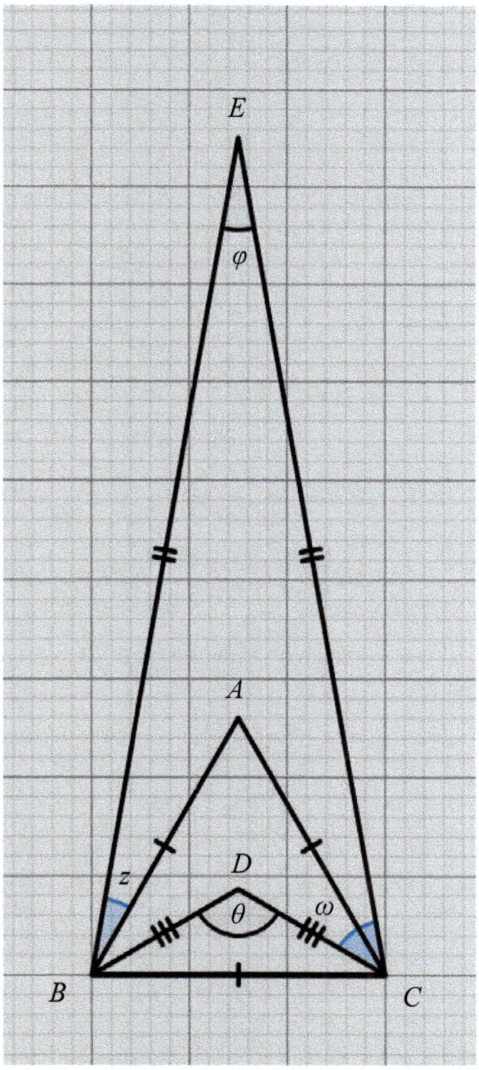

An equilateral triangle *ABC,* an isosceles triangle *DBC* and another isosceles triangle *EBC* share the same side *BC*. Also, angle $\measuredangle BDC = \theta = 120^0$ and $\measuredangle BEC = \varphi = 20^0$.

Find the shaded blue angles ω and z

Angles in Polygons

EXAM PRACTICE QUESTION 4.40 (ANSWERS: (a) $y = 120^0$, $x = 60^0$, (b) $y = x = 90^0$, (c) $y = 72^0$, $x = 108^0$, (d) $y = 60^0$, $x = 120^0$, (e) $y = \left(51\frac{3}{7}\right)^0$, $x = \left(128\frac{4}{7}\right)^0$, (f) $y = 45^0$, $x = 135^0$, (g) $y = 40^0$, $x = 140^0$, (h) $y = 36^0$, $x = 144^0$, (i) $y = \left(38\frac{8}{11}\right)^0$, $x = \left(141\frac{3}{11}\right)^0$, (j) $y = 30^0$, $x = 150^0$, (k) $y = 18^0$, $x = 162^0$,)

Find the interior and exterior angles x and y respectively for the following regular polygons:

(a) Equilateral triangle, (b) Square, (c) Pentagon, (d) Hexagon, (e) Heptagon, (f) Octagon, (g) Nonagon, (h) Decagon, (i) Endecagon, (j) Dodecagon, (k) Eicosagon.

EXAM PRACTICE QUESTION 4.41 (ANSWER: $x = 162^0$)

Find the interior angle of a regular eicosagon using the formula $20x = (n-2)\times 180°$, where x denotes the interior angle.

EXAM PRACTICE QUESTION 4.42 (ANSWER: $\varphi = 72^0$, $\theta = 102^0$, $\omega = 90^0$)

On each different side of an equilateral triangle draw a square, a regular pentagon and a regular hexagon, as shown in the following figure. Find the measures of the shaded blue angles, between the different shapes.

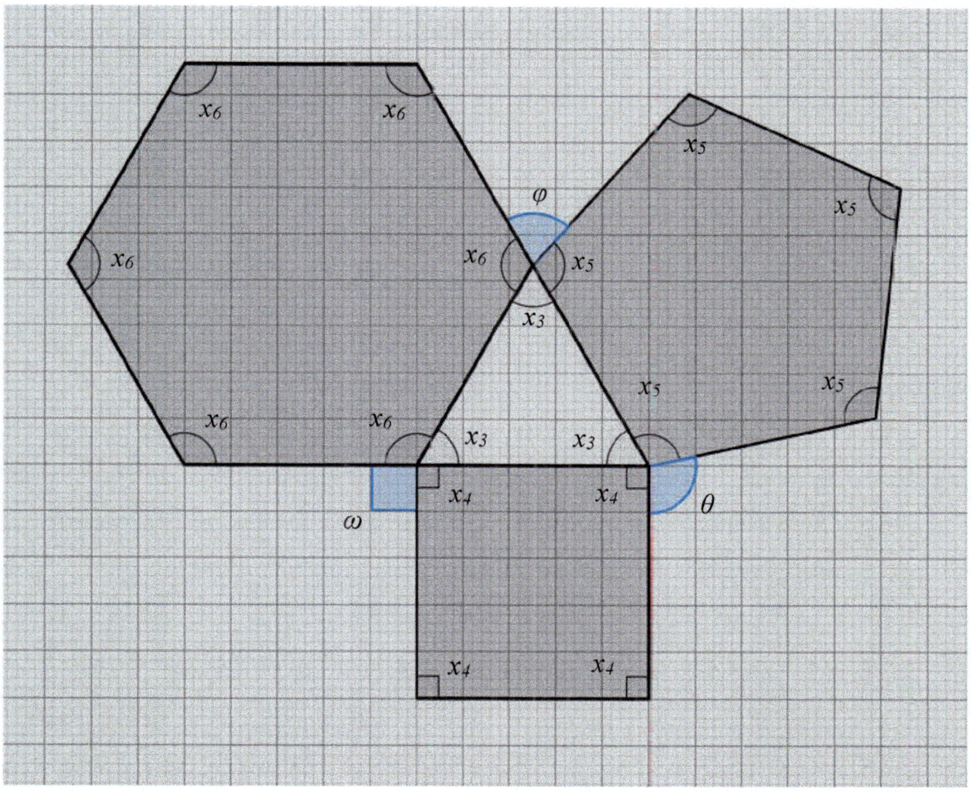

Topic 4.3 QUADRILATERALS (SQUARE, RECTANGLE, PARALLELOGRAM, TRAPEZIUM, KITE, RHOMBUS, ARROW).

Write the formula for the area, adjacent to each one of the following shapes:

Square

Area of square =

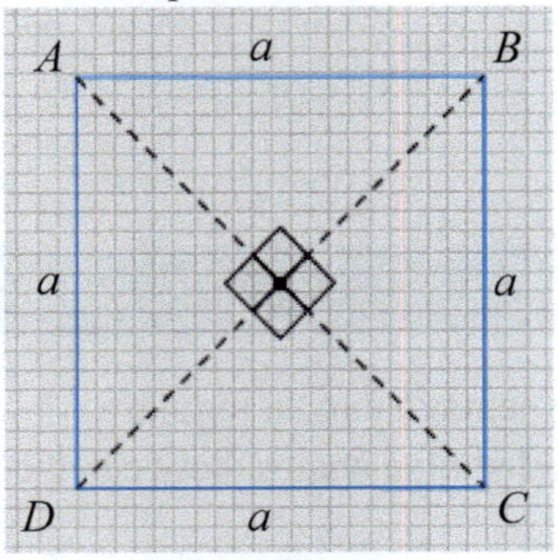

Rectangle

Area of Rectangle =

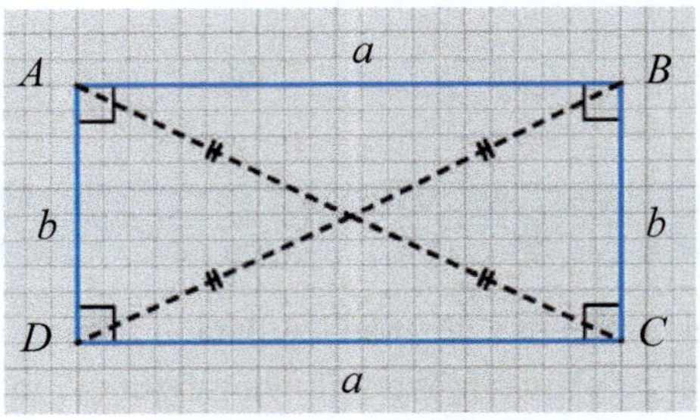

Isosceles Trapezium

Area of isosceles trapezium =

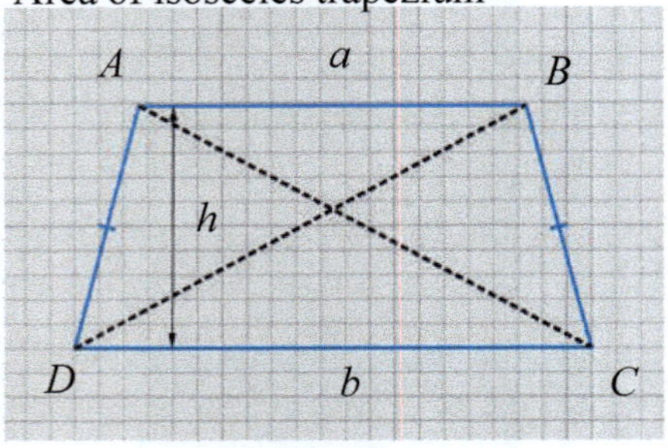

Trapezium

Area of trapezium =

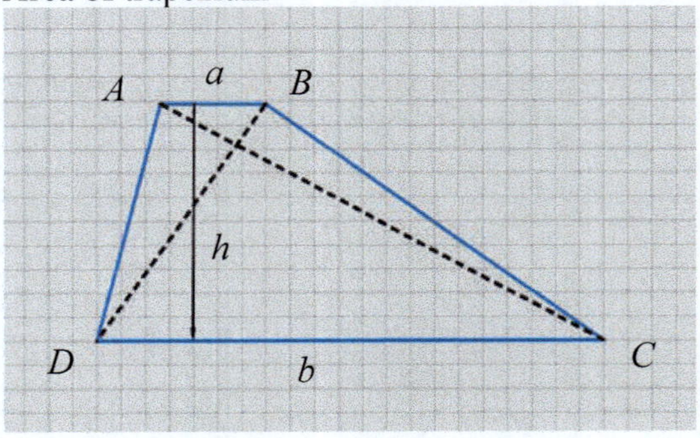

Trapezium

Area of trepazium =

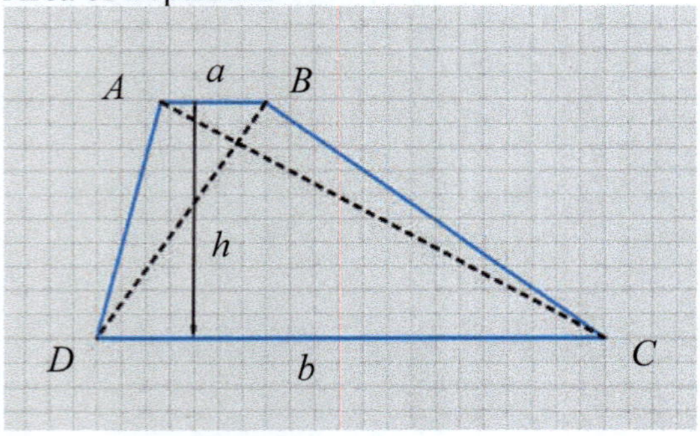

Rhombus

Area of Rhombus =

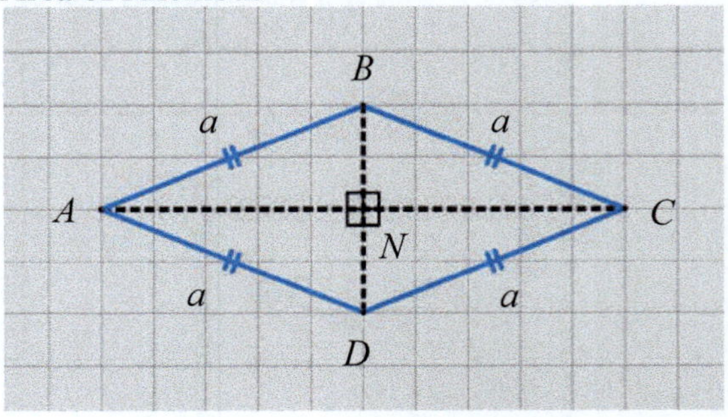

Kite

Area of kite =

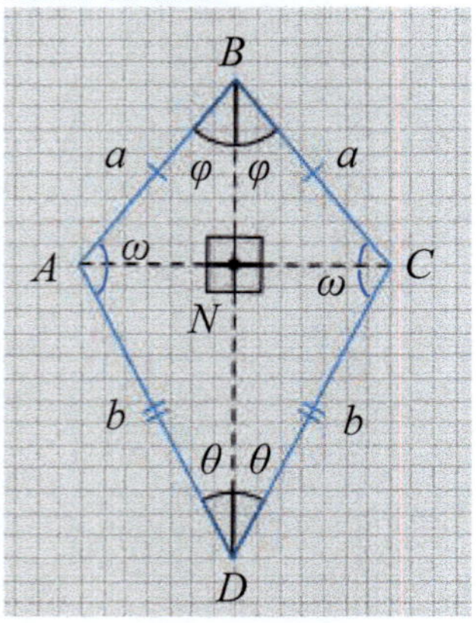

Arrow

Area of arrow =

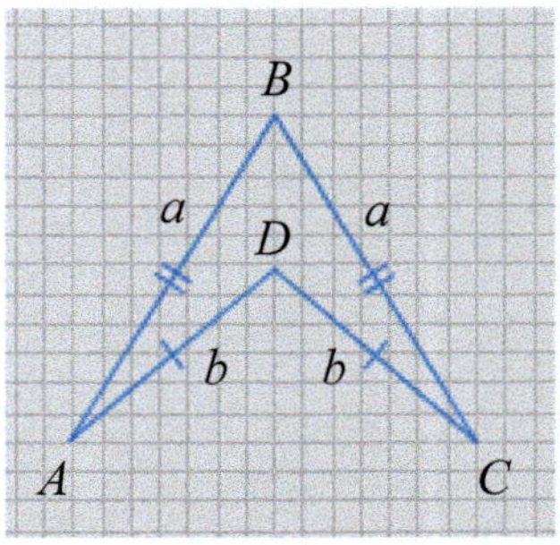

Topic 4.4 CONSTRUCTION AND LOCI.

Exam Practice Question 4.43

Construct a scalene triangle of sides 3.5 cm, 4 cm and 5 cm.

Exam Practice Question 4.44

Construct a scalene triangle of sides 7 cm, 8 cm and 10 cm.

Exam Practice Question 4.45

Construct an equilateral triangle with sides of 4 cm.

Exam Practice Question 4.46

Draw a horizontal line segment AB equal to 10 cm and consider a point C which divides it in the ratio 4 : 6 ($AC : CB$). Construct a perpendicular line through C. Explain your steps.

Exam Practice Question 4.47

Draw two intersecting lines AB and BC making an angle of 60°.
Find the locus of the points which are always on equal distance from each one of the lines.

EXAM PRACTICE QUESTION 4.48

Find the locus of all the points equidistant from a fixed line XY, by distance d.

EXAM PRACTICE QUESTION 4.49

Shade the region of points which are in a distance equal or less than 2 cm from a fixed point C.

EXAM PRACTICE QUESTION 4.50

Mark two points A and B, 5 cm apart.
Shade the region of all the points between A and B, which are closer to B than to A.

Topic 4.5 CONGRUENT TRIANGLES.

EXAM PRACTICE QUESTION 4.51 **(ANSWERS: (a) Yes, (b) Yes, (c) Yes, (d) No)**

Check whether the triangles in the following figures are congruent. Give reasons

(a)

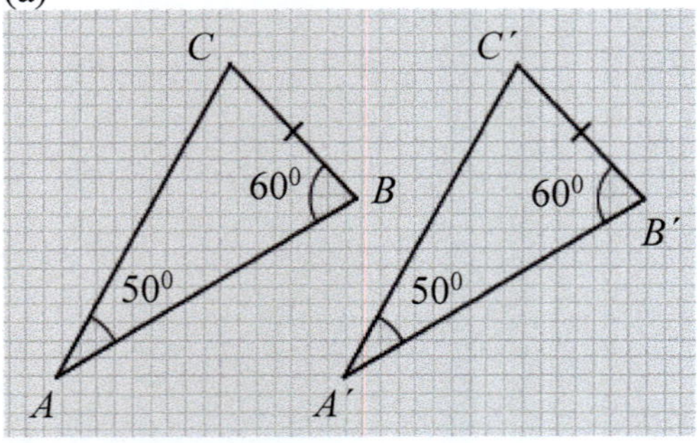

(b)

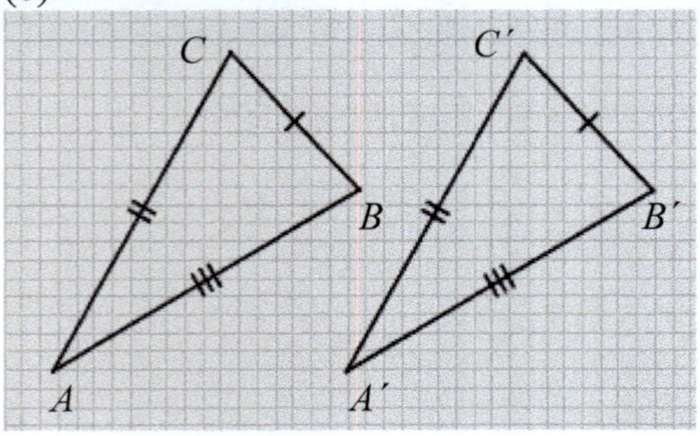

(c)

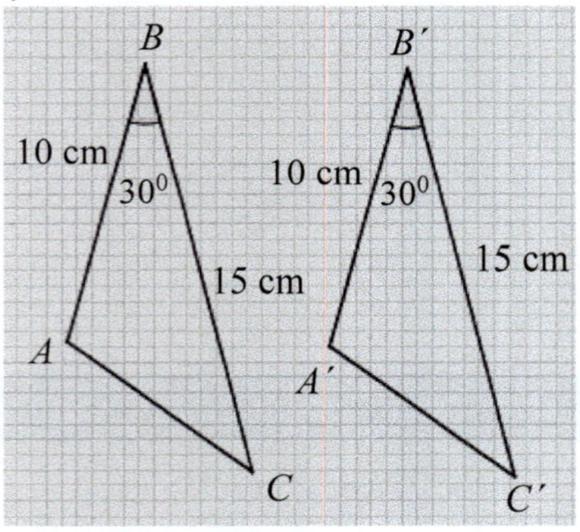

(d)

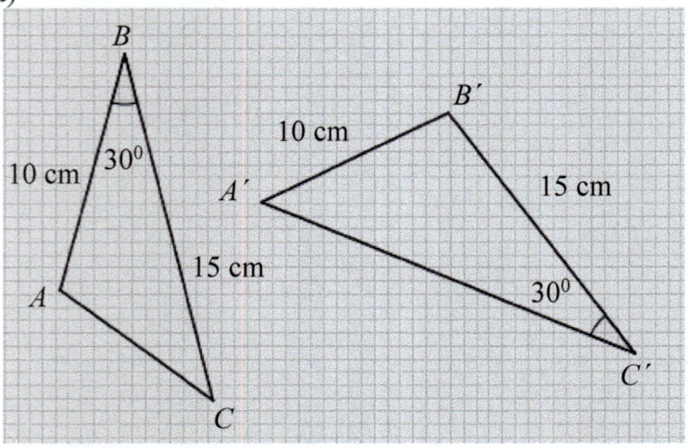

Topic 4.6 SIMILAR TRIANGLES.

EXAM PRACTICE QUESTION 4.52 (ANSWER: YES, the triangles are similar)

Are the triangles *ABC* and *DEF* similar?

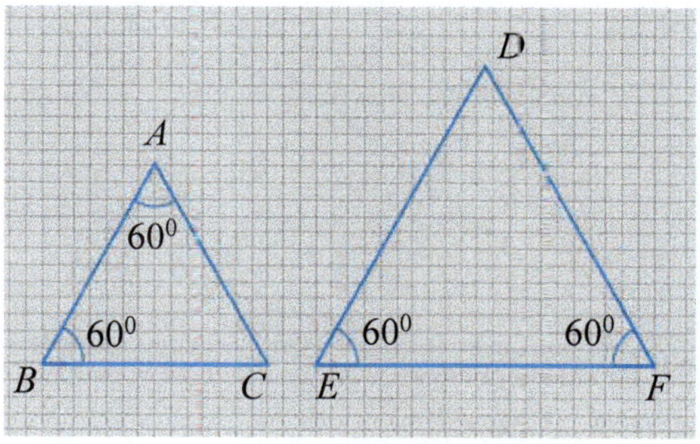

EXAM PRACTICE QUESTION 4.53 (ANSWERS: (ii) $\hat{A} = \hat{D} = 90^0$, $\hat{C} = \hat{E} = \varphi$, (iii) $\frac{AB}{DF} = \frac{BC}{EF} = \frac{AC}{DE}$)

The right-angled triangles *ABC* and *DEF* have:

$$B\hat{C}A = F\hat{E}D = \varphi$$

(i) Show that the two triangles *ABC* and *FED* are similar.

(ii) Write down the angles which are equal.

(iii) Write down the ratios of the sides which are in proportion.

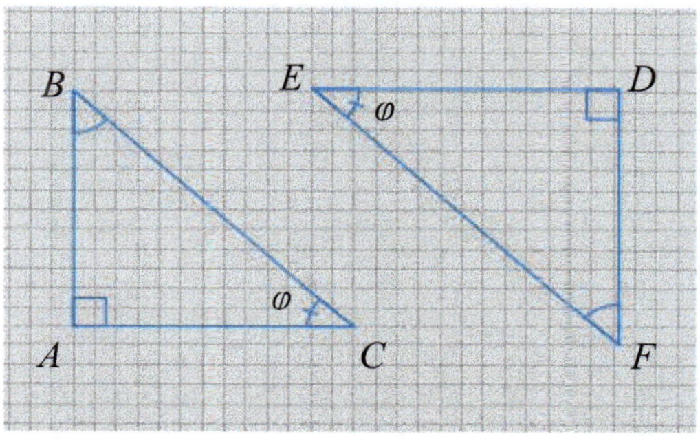

EXAM PRACTICE QUESTION 4.54

(ANSWERS: (a) $\frac{AB}{AD} = \frac{AC}{AE} = \frac{BC}{DE} = \frac{2}{3}$), (b-i) $y = 3.5$, (b-ii) $AE = 7.5$

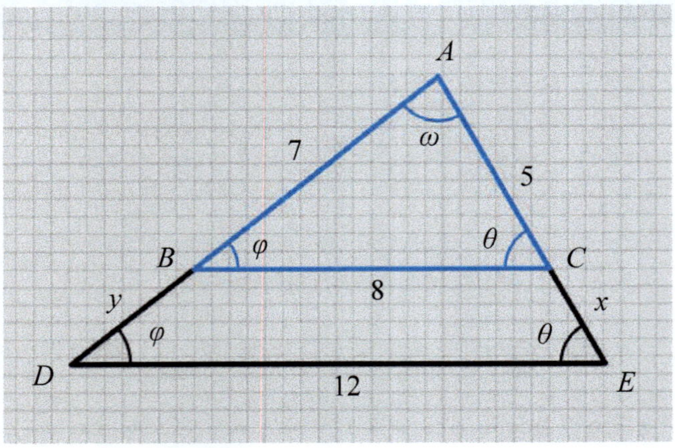

Referring to the adjacent figure:

(a) Prove that the triangles ABC and ADE are similar triangles and determine their similarity ratio.

(b) Determine (i) y and (ii) AE.

EXAM PRACTICE QUESTION 4.55

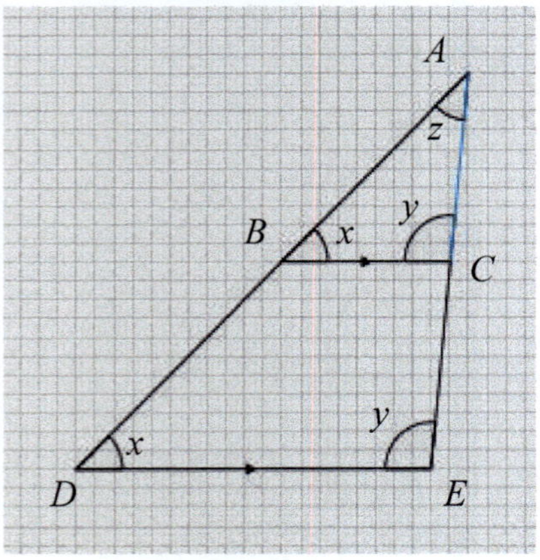

In the adjacent figure, BC is parallel to DE. Prove that the triangles ABC and ADE are similar.

EXAM PRACTICE QUESTION 4.56 (ANSWER: $x = 4.6$ m)

In the Figure below, $DE = 1.5$ m is parallel to $BC = 3$ m and $AD = 4.6$ m. Workout the value of x.

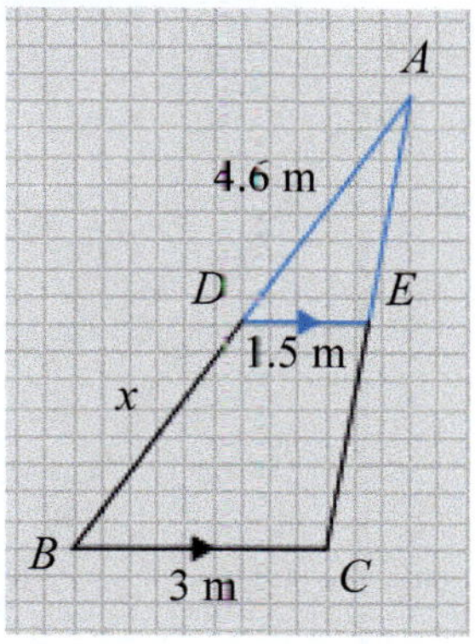

EXAM PRACTICE QUESTION 4.57 (ANSWER: $x = 7.2$ cm)

The adjacent figure shows lines AE and BD intersecting at C, with AB parallel to DE. Also, $AC = 10$ cm, $CB = 12$ cm and $CE = 6$ cm. Show that the triangles ABC and EDC are similar. Find the distance x.

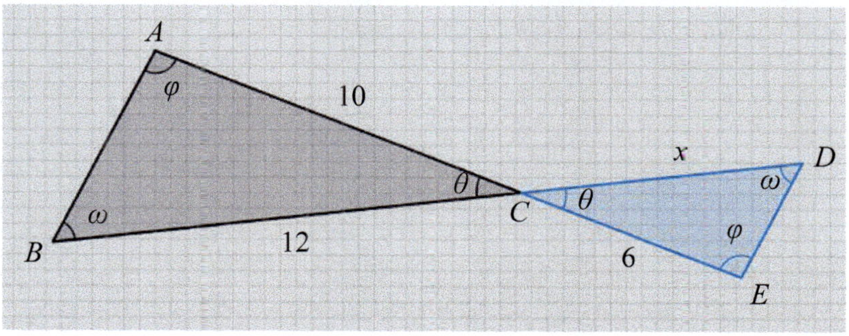

EXAM PRACTICE QUESTION 4.58

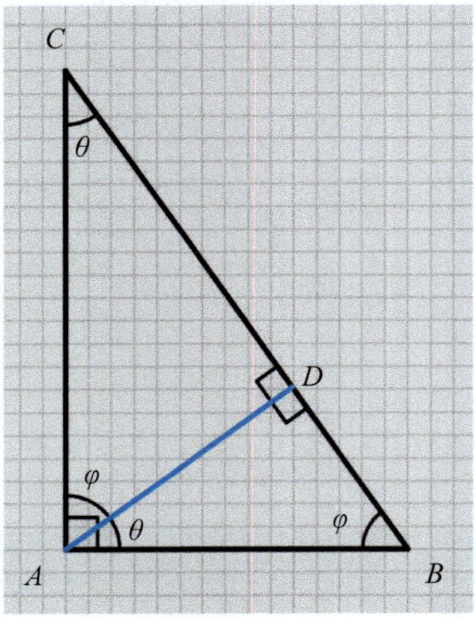

Consider the right-angled triangle in the adjacent figure, with $B\hat{A}C = 90°$ and its height AD.

(i) Show that the triangles DBA and ABC are similar.

(ii) Also show that the triangles DCA and ABC are similar as well.

(iii) Hence show that $\frac{BC}{AB} = \frac{AB}{DB}$ and $\frac{BC}{AC} = \frac{AC}{DC}$

(iv) Combine the above and prove that $AB^2 + AC^2 = BC^2$ (Pythagoras' Theorem)

EXAM PRACTICE QUESTION 4.59 (ANSWER: $\frac{h_2}{h_1} = 2$)

Two similar scalene triangles have areas 40 cm^2 and 10 cm^2. Determine the ratio of their corresponding heights.

EXAM PRACTICE QUESTION 4.60

(ANSWERS: (a) $AB = 6$ cm, (b) $\dfrac{DC}{EB} = \dfrac{5}{4}$, (c) $\dfrac{\text{Area } \Delta ADC}{\text{Area } \Delta AEB} = \dfrac{25}{16}$, (d) $\dfrac{\text{Area } DEBC}{\text{Area } \Delta ADC} = \dfrac{9}{25}$)

In the adjacent figure, $EB \parallel DC$, $DE = 2$ cm, $AE = 8$ cm and $BC = 1.5$ cm.
Determine

(a) AB

(b) $\dfrac{DC}{EB}$

(c) $\dfrac{\text{Area } \Delta ADC}{\text{Area } \Delta AEB}$

(d) $\dfrac{\text{Area } DEBC}{\text{Area } \Delta ADC}$

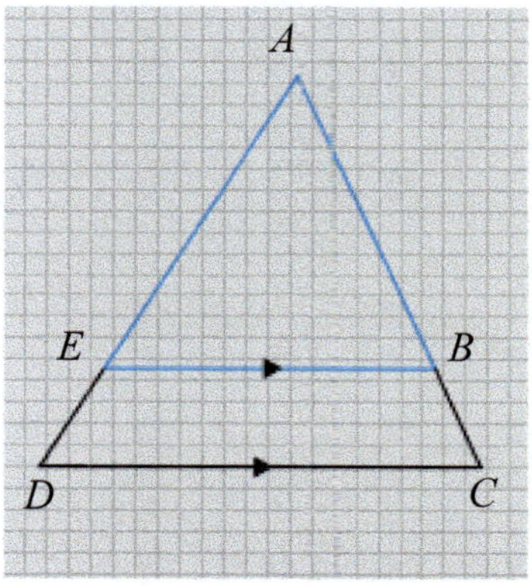

EXAM PRACTICE QUESTION 4.61

(ANSWERS: (i) $DC = 4$ cm, (ii) $\dfrac{\text{Area } \Delta EDC}{\text{Area } \Delta ABC} = \dfrac{1}{9}$)

In the adjacent figure $AB = 15$ cm is parallel to $ED = 5$ cm.
Also, $BC = 12$ cm
Find:

(i) DC

(ii) $\dfrac{\text{Area } \Delta EDC}{\text{Area } \Delta ABC}$

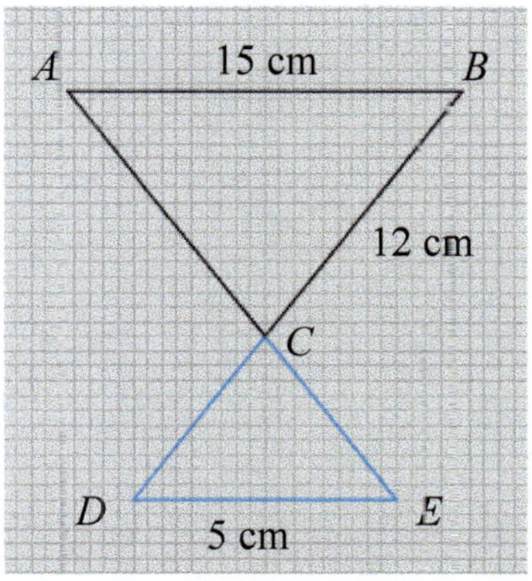

EXAM PRACTICE QUESTION 4.62

(ANSWER: $\dfrac{\text{Area } ABCD}{\text{Area } A'B'C'D'} = \dfrac{4}{9}$)

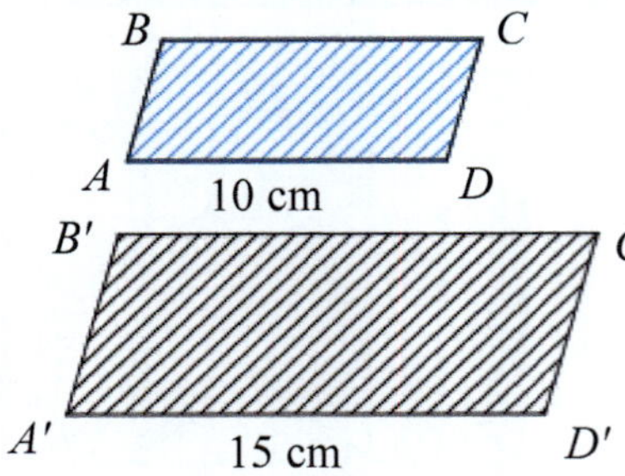

The parallelograms $ABCD$ and $A'B'C'D'$ are similar.
Find the ratio of the areas $ABCD$ to $A'B'C'D'$, when $AD = 10$ cm and $A'D' = 15$ cm.

EXAM PRACTICE QUESTION 4.63

(ANSWERS: (i) $D(16,-10)$, $E(30,4)$, $F(24,-10)$, (ii) $BC = \sqrt{58}$)

Plot the triangle ABC with vertices $A(8,-5)$, $B(15, 2)$ and $C(12, -5)$.

(i) Then enlarge the triangle ABC by a scale factor 2 about the origin. State the coordinates of D, E and F of the image triangle by drawing constructional lines to locate the new positions of D, E and F.

(ii) Determine the length BC in surd form and verify that the corresponding side of the image is twice this length.

(iii) Show that the image is a similar triangle to the object.

EXAM PRACTICE QUESTION 4.64 (ANSWERS: (i) $AD = 6$ cm, (ii) $E(10,6)$)

In the adjacent figure, it is given that:
(a) the coordinates of A are (0, 0),
(b) $\hat{A} = \hat{D} = 90^0$,
(c) $\hat{B} = 45^0$,
(d) $AB = 16$ cm,
(e) $\dfrac{\text{Area } \Delta ABC}{\text{Area } \Delta DEC} = \dfrac{64}{25}$.

Determine:
(i) The distance between A and D
(ii) The coordinates of E

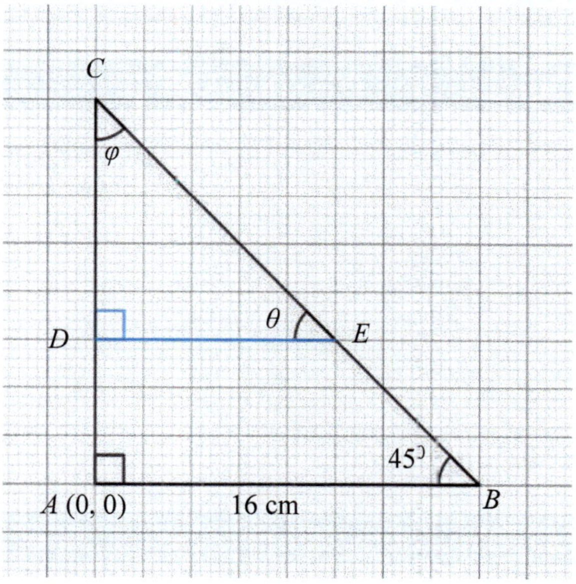

Topic 4.7 PYTHAGORAS' THEOREM.

EXAM PRACTICE QUESTION 4.65 **(ANSWER: $\alpha = 13$)**

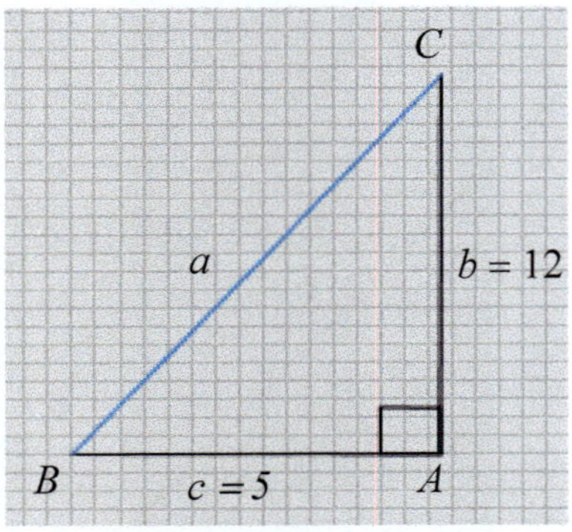

Consider the right-angled triangle *ABC* in the adjacent figure and workout the hypotenuse *BC*.

EXAM PRACTICE QUESTION 4.66 **(ANSWER: $AB = 5$)**

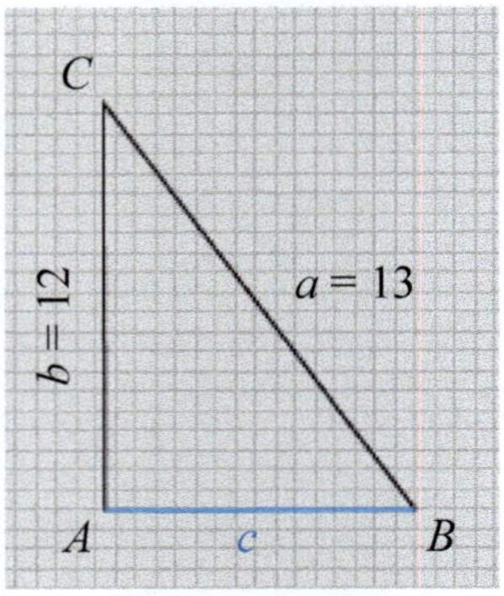

Given that $BC = a = 13$, and $AC = b = 12$ in the adjacent figure, workout the side *AB*.

EXAM PRACTICE QUESTION 4.67 **(ANSWER: $x = 12.8$ cm to 3 s.f.)**

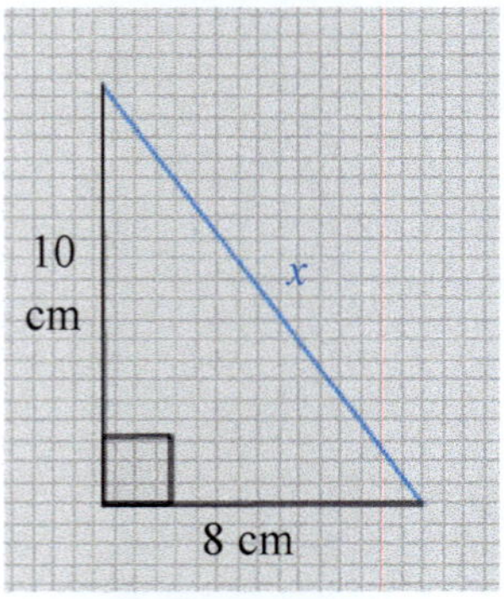

In the adjacent figure, find the size of the hypotenuse, x.

EXAM PRACTICE QUESTION 4.68 **(ANSWER: $y = 11.6$ cm to 3 s.f.)**

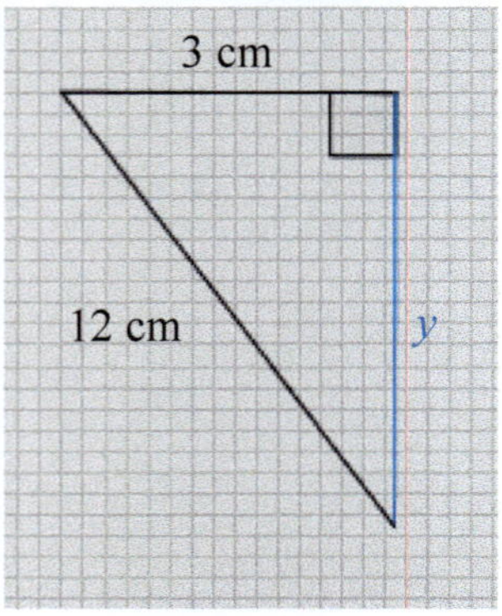

Workout the side y in the right-angled triangle of the adjacent figure.

EXAM PRACTICE QUESTION 4.69 (ANSWER: $h = 2\sqrt{3}$ m)

A ladder of length 4 m leans against a wall so that the foot of the ladder is 2 m from the base of the wall. Find the height of the wall where the ladder rests, in surd form.

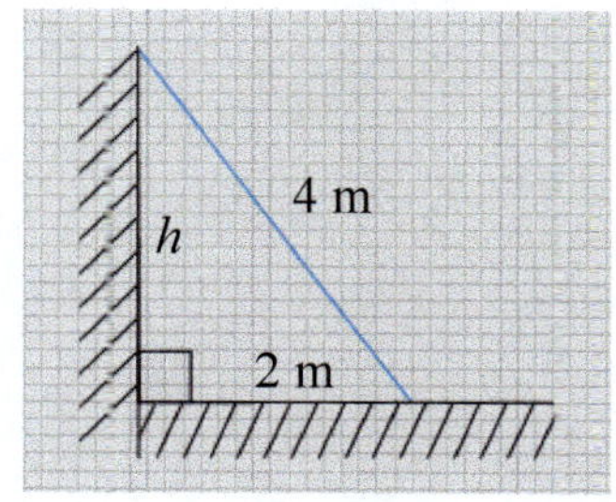

EXAM PRACTICE QUESTION 4.70 (ANSWERS: (i) 6.32, (ii) 8, (iii) 4.47 to 3 s.f., (iv) 6, (v) 5.39 to 3 s.f.)

Plot the following sets of points: (i) $A(2, 2)$, $B(8, 4)$, (ii) $A(0, 0)$, $B(0, 8)$, (iii) $A(-2, 4)$, $B(0, 8)$, (iv) $A(2, 8)$, $B(8, 8)$, (v) $A(1, 3)$, $B(6, 1)$. Determine the lengths AB in each case to 3 s.f.

EXAM PRACTICE QUESTION 4.71 **(ANSWERS: $B(1,3)$, $C(2,2)$, $D(3,1)$, $E(4,3)$, $F(5,2)$, $G(6,3)$, $H(6,1)$, (i) 3.61, (ii) 3.61, (iii) 4.12,(iv) 2.83, (v) 5.39, (vi) 4.12, (vii) 5, (viii) 3, (ix) 5, (x) 5.39)**

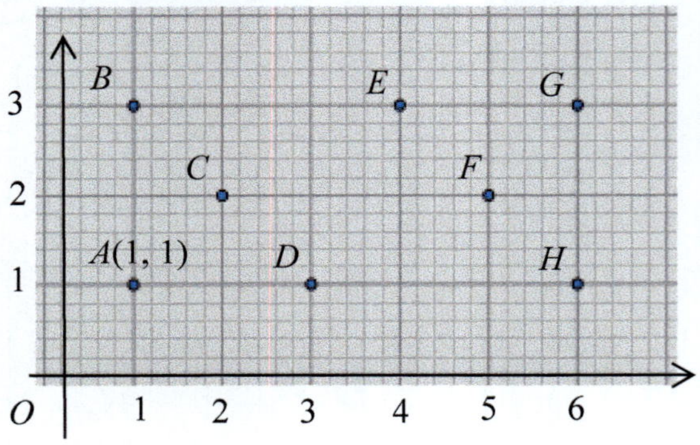

Given that A's coordinates are (1, 1), state the coordinates of the points B, C, D, E, F, G and H, in the adjacent figure.

Then, determine the following lengths:

(i) AE (ii) DG (iii) CH (iv) BD
(v) BH (vi) AF (vii) BG (viii) CF
(ix) AH (x) AG.

EXAM PRACTICE QUESTION 4.72 (ANSWER: $C = 16\text{ m}^2$)

In the adjacent figure, find the area *C*.

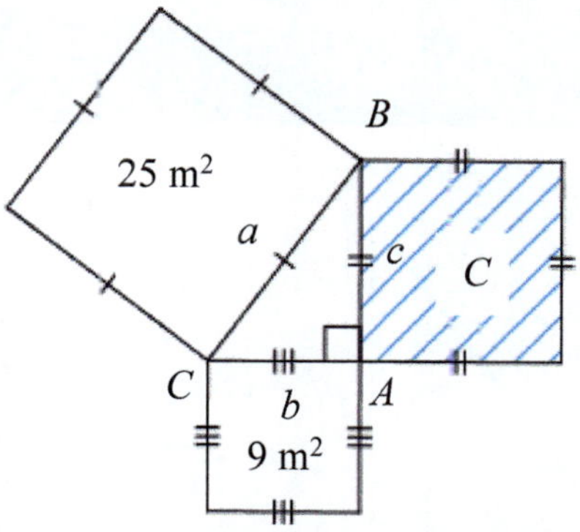

EXAM PRACTICE QUESTION 4.73 (ANSWERS: (i) 13 cm, (ii) $A\hat{B}C = 45.2^0$ to 3 s.f., $B\hat{C}D = 134.8^0$ to 4 s.f.)

The diagonals *AC* and *BD* of the rhombus *ABCD* are 10 cm and 24 cm respectively. Find

(i) the length of each side

(ii) the angles $A\hat{B}C$ and $B\hat{C}D$.

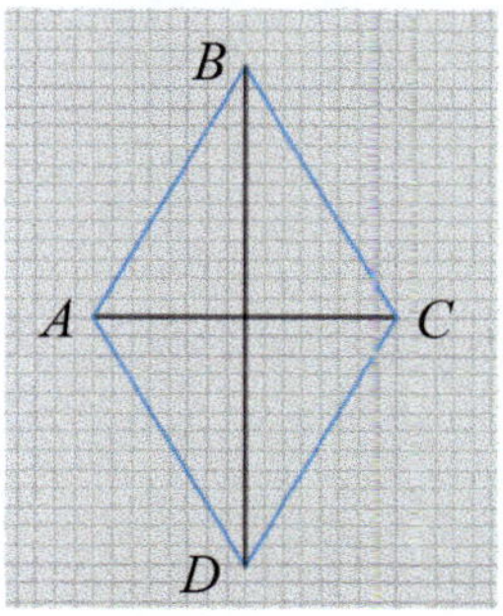

EXAM PRACTICE QUESTION 4.74 (ANSWER: 6.61 cm to 3 s.f.)

A chord *AB* of length 15 cm is drawn in a circle of radius 10 cm. Find the shortest length from the centre to the chord.

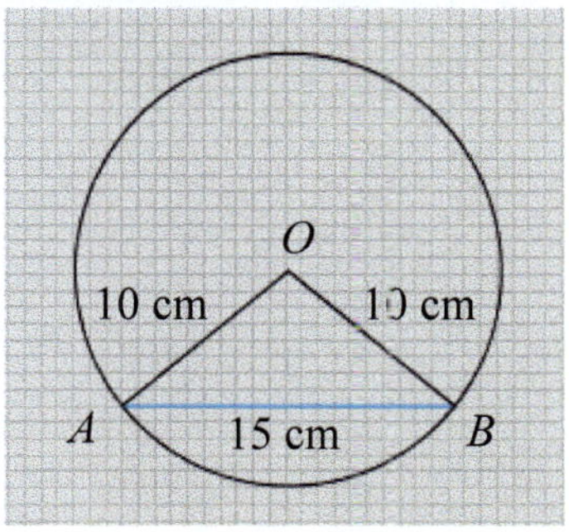

Pythagorean Triples

EXAM PRACTICE QUESTION 4.75 **(ANSWERS: (i) $a^2 = b^2 + c^2$, (ii) $P = 2x(x+y)$, (iii) $A = x^3y - xy^3$)**

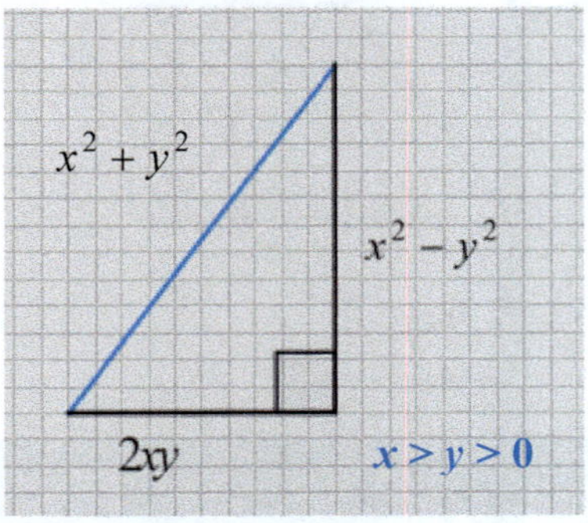

(i) Prove that the integers $a = x^2 + y^2$, $b = x^2 - y^2$ and $c = 2xy$ where $x > y > 0$ are two arbitrary positive integers, form a Pythagorean triple.

(ii) Calculate the perimeter of the triangle with sides the above integers *a*, *b* and *c*.

(iii) Calculate the area of the triangle with sides the above integers *b* and *c*.

EXAM PRACTICE QUESTION 4.76 **(ANSWERS: (i) $a^2 = b^2 + c^2$, (ii) $P = 4x^2 + 6x + 2$, (iii) $A = x(x+1)(2x+1)$)**

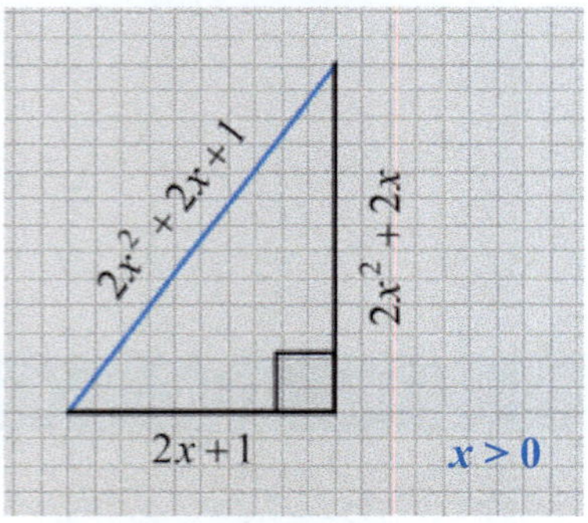

(i) Prove that the integers $a = 2x^2 + 2x + 1$, $b = 2x^2 + 2x$ and $c = 2x + 1$, where $x > 0$ is an arbitrary positive integer, form a Pythagorean triple.

(ii) Calculate the perimeter of the triangle with sides the above integers *a*, *b* and *c*.

(iii) Calculate the area of the triangle with sides the above integers *a*, *b* and *c*.

EXAM PRACTICE QUESTION 4.77 (ANSWER: $y = 2x^2 + 2x$)

In a right-angled triangle, the hypotenuse is given by the expression $2x^2 + 2x + 1$ and one of the vertical sides is $2x + 1$. Find the other vertical side in terms of x.

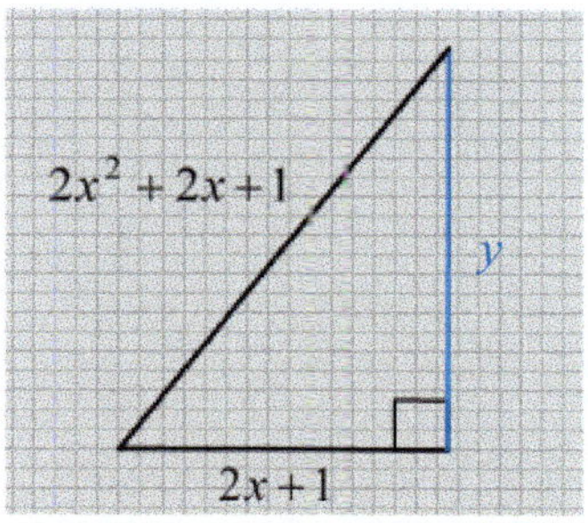

EXAM PRACTICE QUESTION 4.78 (ANSWERS: (i) $x = 2y^2 + 2y + 1$, (ii) $P = 4y^2 + 6y + 2$, (iii) $A = 2y^3 + 3y^2 + y$)

If the vertical sides of a right – angled triangle are given as $2y^2 + 2y$ and $2y + 1$, determine:

(i) the hypotenuse in terms of y.
(ii) the perimeter of the triangle.
(iii) the area of the triangle.

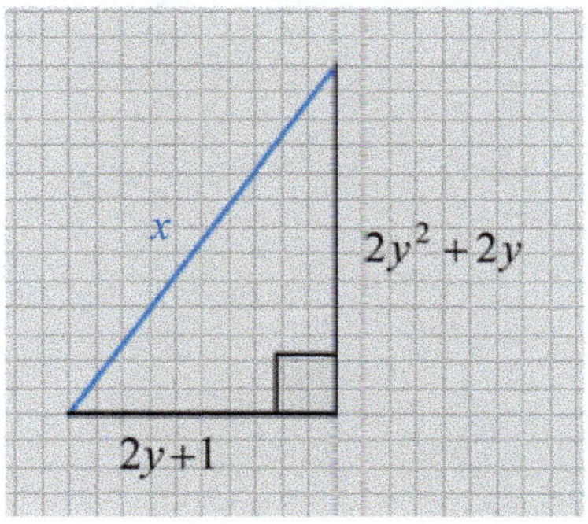

EXAM PRACTICE QUESTION 4.79 (ANSWERS: (i) $x = 2y^2 + 2y + 1$, (ii) 5, 12, 13, (iii) 7, 24, 25)

The vertical sides of a right – angled triangle are given by the expressions $2y + 1$ and $2y^2 + 2y$.

(i) Determine the hypotenuse in terms of y
(ii) If $y = 2$, find the three sides.
(iii) If $y = 3$, find the three sides.

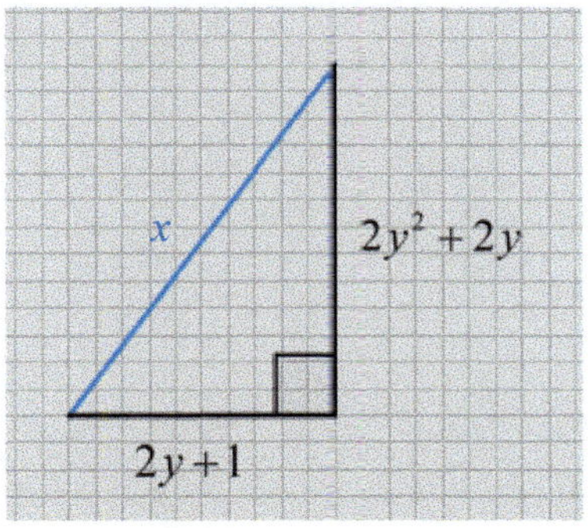

EXAM PRACTICE QUESTION 4.80

Verify that the following are Pythagorean triples

(i) (3, 4, 5) (ii) (5, 12, 13) (iii) (7, 24, 25), (iv) (9, 40, 41) (v) (11, 60, 61)

Draw the corresponding triangles.

Topic 4.8 TRIGONOMETRIC RATIOS LENGTHS AND ANGLES IN THE RIGHT ANGLED TRIANGLES.

EXAM PRACTICE QUESTION 4.81 (ANSWERS: (i) – 0.866, (ii) – 1, (iii) 0.866)

Use the unit radius circle to find:

(i) sin 240° (ii) tan 135° (iii) cos 330°.

Use the trigonometric exact values, not a calculator.

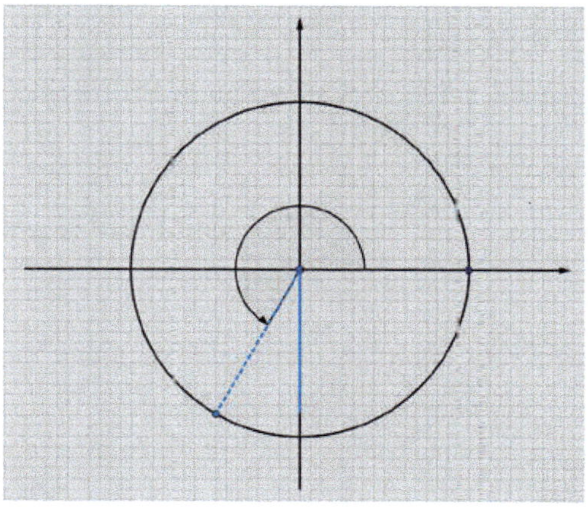

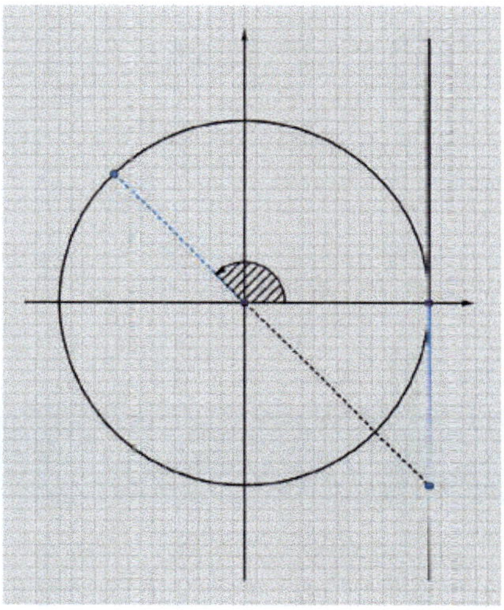

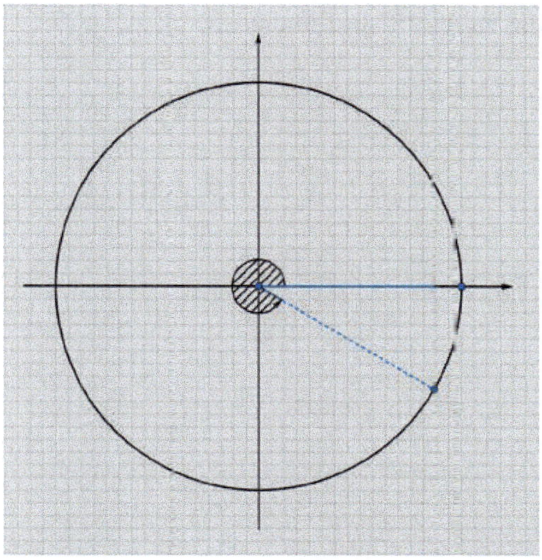

EXAM PRACTICE QUESTION 4.82

(ANSWERS: (a) 24.6^0 to 3 s.f., (b) 73.0^0 to 3 s.f., (c) 53.1^0 to 3 s.f., (d) 36.9^0 to 3 s.f.)

Using the given lengths, work out the size of the angle x to 3 s.f. for the following right-angled triangles (use your calculator).

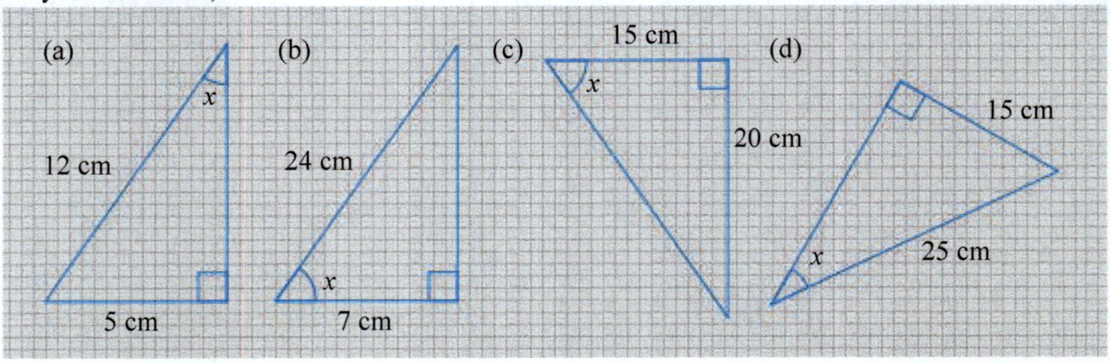

Triangles are drawn not to scale

EXAM PRACTICE QUESTION 4.83

(ANSWERS: $\sin x = \frac{4}{5}$, $\cos x = \frac{3}{5}$)

If $\tan x = \frac{4}{3}$ sketch a right – angled triangle, hand mark the sides 3 and 4, hence find the third side. What are the values of $\sin x$ and $\cos x$.

EXAM PRACTICE QUESTION 4.84

(ANSWERS: the third side is 5, $\cos x = \frac{5}{13}$, $\tan x = \frac{12}{5}$)

If $\sin x = \frac{12}{13}$ sketch a right – angled triangle ABC and mark the sides 12 and 13. Hence find the third side. What are the values of cos x and tan x ?

EXAM PRACTICE QUESTION 4.85

(ANSWERS: (a) $\sin x = \frac{7 \times \sqrt{130}}{130}$, $\cos x = \frac{9 \times \sqrt{130}}{130}$, $\tan x = \frac{7}{9}$, (b) $\sin y = \frac{9 \times \sqrt{130}}{130}$, $\cos y = \frac{7 \times \sqrt{130}}{130}$, $\tan y = \frac{9}{7}$, (c) $x = 37.9$ to 3 s.f., $y = 52.1$ to 3 s.f.)

In the right – angled triangle of the figure below, find the following trigonometric ratios:

(a) sin x, cos x, tan x

(b) sin y, cos y, tan y, giving your answers in surd form

(c) calculate the angles x and y to 3 s.f., using your calculator.

EXAM PRACTICE QUESTION 4.86 (ANSWERS: (i) 0.342, (ii) – 0.819, (iii) 28.6)

Given that sin 20° = 0.342, cos 35° = 0.819, and tan 88° = 28.6, find the values:
(i) sin 160° (ii) cos 145° (iii) tan 268°. Do not use calculator.

EXAM PRACTICE QUESTION 4.87 (ANSWER: $\tan 120^0 = -\sqrt{3}$, $\tan 135^0 = -1$, $\tan 180^0 = 0$)

Write down the values of the tangent for the following angles: 120°, 135°, 180°.

EXAM PRACTICE QUESTION 4.88

Use a calculator to find the values of the sine, cosine and tangent of the following angles:
15°, 25°, 32°, 65°, 88°, 93°, 99°, 136°, 178°, 200°, 320°.

EXAM PRACTICE QUESTION 4.89

(ANSWERS: (i) 11.03 to 2 d.p., (ii) 0.80 to 2 d.p., (iii) – 39.71 to 2 d.p.)

Evaluate the following with a calculator to 2 d.p.

(i) 3 sin 25° + 5 cos 35° + 7 tan 39°

(ii) cos 35° – sin 79° + tan 44°

(iii) 5 cos 35° + 6 sin 90° –7 tan 82°.

EXAM PRACTICE QUESTION 4.90

(ANSWERS: (i) 30^0, 150^0, (ii) 150^0, 210^0, (iii) 45^0, 225^0, (iv) 240^0, 300^0, (v) 300^0, 120^0)

Find all the values of x between 0° and 360°, if

(i) $\sin x = \frac{1}{2}$, (ii) $\cos x = -\,0.866$, (iii) $\tan x = 1$, (iv) $\sin x = -\frac{\sqrt{3}}{2}$, (v) $\tan x = -\,\sqrt{3}$.

EXAM PRACTICE QUESTION 4.91 (ANSWERS: (i) $\cos\varphi = \frac{\sqrt{33}}{7}$, $\tan\varphi = \frac{4\sqrt{33}}{33}$

(ii) $\cos\varphi = -\frac{\sqrt{33}}{7}$, $\tan\varphi = -\frac{4\sqrt{33}}{33}$)

If $\sin\varphi = \frac{4}{7}$ find cos φ and tan φ, without the use of a calculator,

(i) if φ is an acute angle, (ii) if φ is an obtuse angle.

EXAM PRACTICE QUESTION 4.92 **(ANSWERS: (i) $x = 30^0, 150^0, -210^0, -330^0$, (ii) $x = 60^0, 120^0, -240^0, -300^0$, (iii) $x = 150^0, 210^0, -150^0, -210^0$, (iv) $x = 30^0, 330^0, -30^0, -330^0$,)**

Use the waveforms of sin x and cos x for $-360° \leq x \leq 360°$ to find the angles for which:

(i) $\sin x = \dfrac{1}{2}$, (ii) $\sin x = \dfrac{\sqrt{2}}{2}$, (iii) $\cos x = -0.866$, (iv) $\cos x = \dfrac{\sqrt{3}}{2}$.

Do not use a calculator.

Topic 4.9 THE AREA OF A TRIANGLE & AREA OF A PARALLELOGRAM, GIVEN THE LENGTHS OF TWO SIDES AND THE ANGLE BETWEEN THEM.

Area of a Triangle given its two sides and the angle between them

EXAM PRACTICE QUESTION 4.93 (ANSWER: 10 cm^2)

Find the area of a triangle, when $a = 8$ cm, $c = 5$ cm and the angle between these sides is 30° .

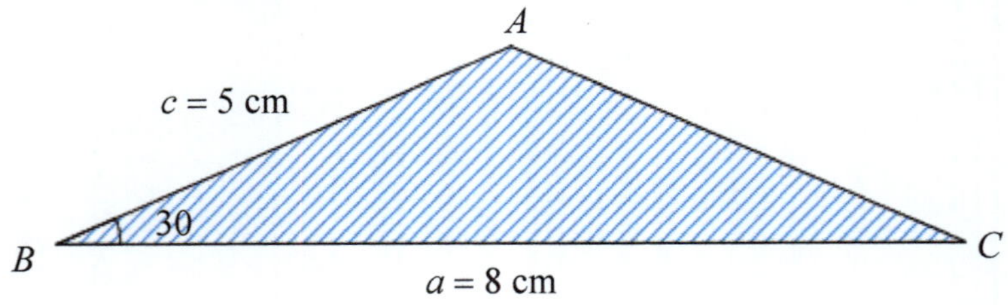

EXAM PRACTICE QUESTION 4.94 (ANSWER: 119.175 cm^2 , 15.6 cm to 3 s.f.)

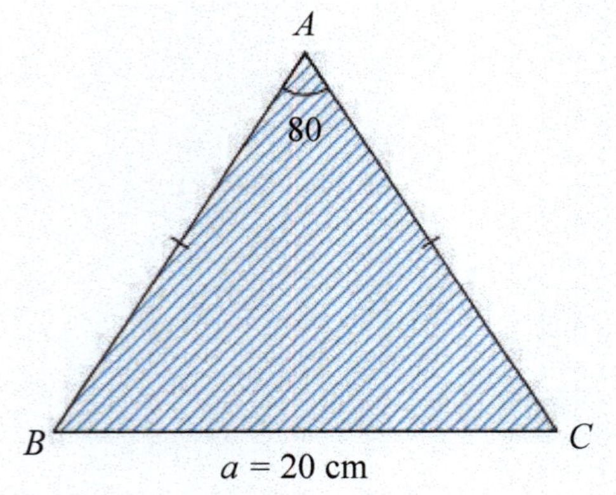

Find the area of the isosceles triangle ABC, of the adjacent figure. Hence find $AB = AC$

EXAM PRACTICE QUESTION 4.95 (ANSWER: 12 cm^2)

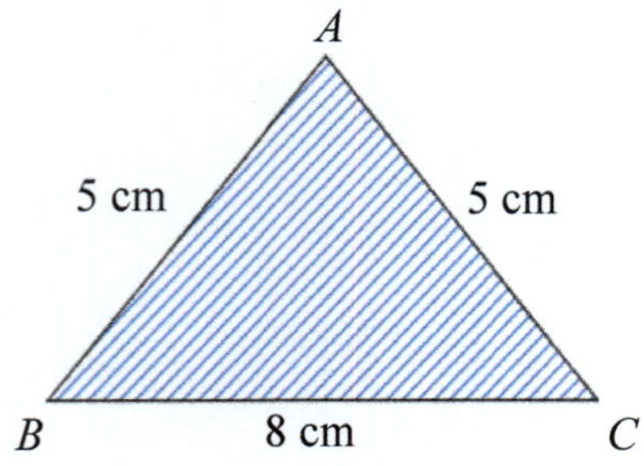

Find the area of the adjacent isosceles triangle.

EXAM PRACTICE QUESTION 4.96 **(ANSWER: 6.25 mm^2)**

In the adjacent figure, find the area of the triangle AOB.

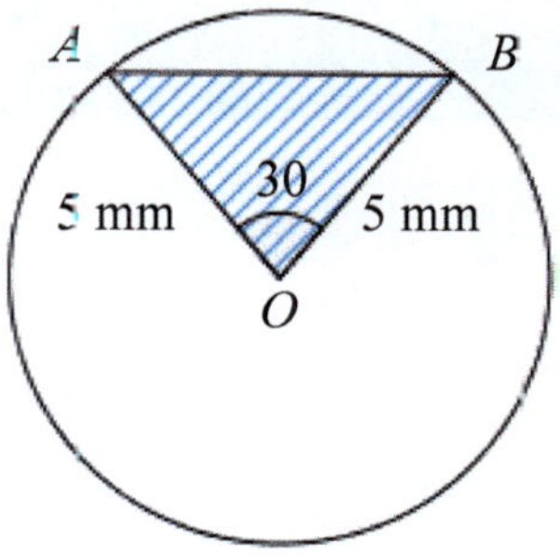

Area of a Parallelogram given its two sides and the angle between them

EXAM PRACTICE QUESTION 4.97 **(ANSWER: 2.5 cm^2)**

As shown in the figure below, the parallelogram $ABCD$ has sides $AB = a = 5$ cm and $BC = b = 2$ cm. Also, its angle DAB is $x = 30^0$. Find the area of the shaded blue triangle ADC.

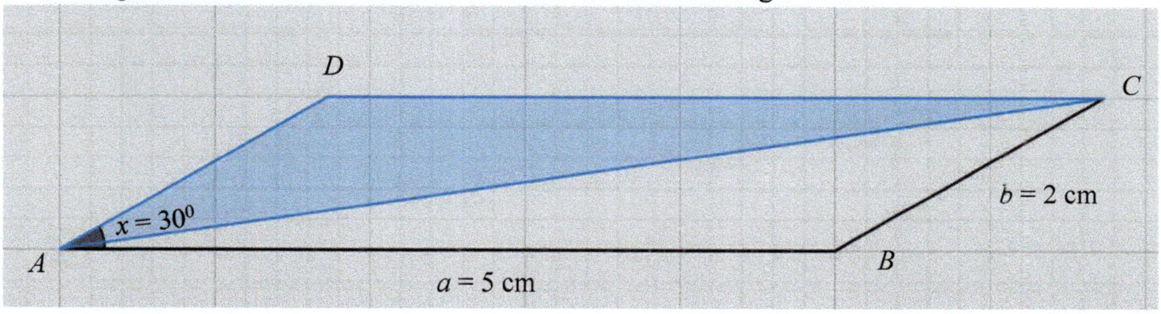

Topic 4.10 THE SINE AND COSINE RULES.

EXAM PRACTICE QUESTION 4.98 **(ANSWERS: (i) $\varphi = 58.1^0$, $\theta = 76.9^0$ or $\varphi = 121.9^0$, $\theta = 13.1^0$, (ii) 13.8 cm or 3.2 cm)**

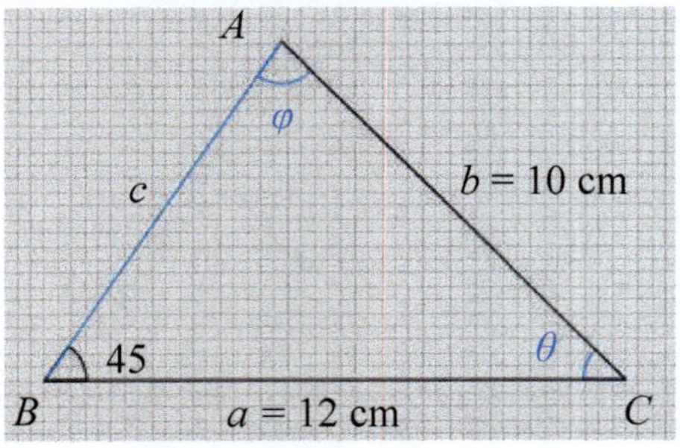

In the adjacent triangle,

(i) Find the values of φ and θ.
(ii) Find the side c.

EXAM PRACTICE QUESTION 4.99 (ANSWER: $c = 28.69$ cm, $\hat{B} = 30.3^0$, $\hat{C} = 105^0$ to 3 s.f.)

The scalene triangle ABC is given with two known sides $a = 21$ cm, $b = 15$ cm and one known angle A = 45^0. Find the third side c and the other two angles (B and C).

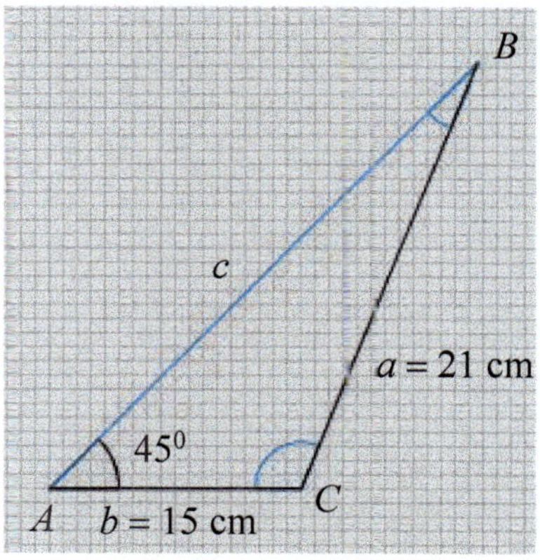

EXAM PRACTICE QUESTION 4.100 (ANSWER: $a = 67.6$ cm to 3 s.f.)

In the following figure, work out the value of the side a.

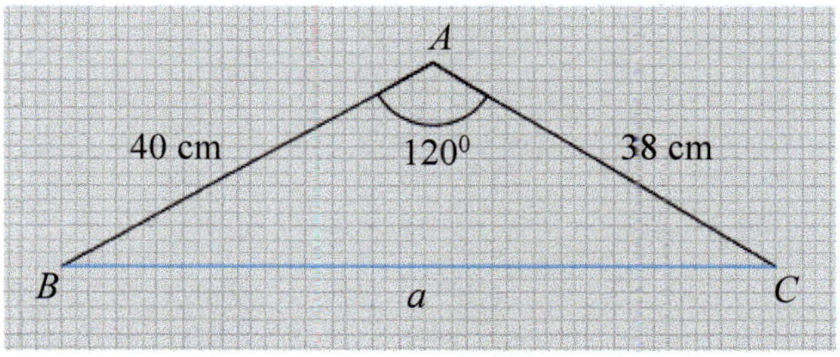

EXAM PRACTICE QUESTION 4.101 (ANSWER: $\hat{A} = 81.8^0$, $\hat{B} = 60.0^0$, $\hat{C} = 38.2^0$, all to 3 s.f.)

In a triangle *ABC*, $b = 7$ cm, $a = 8$ cm and $c = 5$ cm. Use the cosine rule to find the angles.

EXAM PRACTICE QUESTION 4.102 (ANSWER: $\hat{A} = 79.8^0$, $\hat{C} = 59.2^0$, $c = 13.1$ cm to 3 s.f. or $\hat{A} = 100.2^0$, $\hat{C} = 38.8^0$, $c = 9.55$ cm to 3 s.f.)

For the triangle *ABC* of the following figure workout the side *AB* and the angles *A* and *C*.

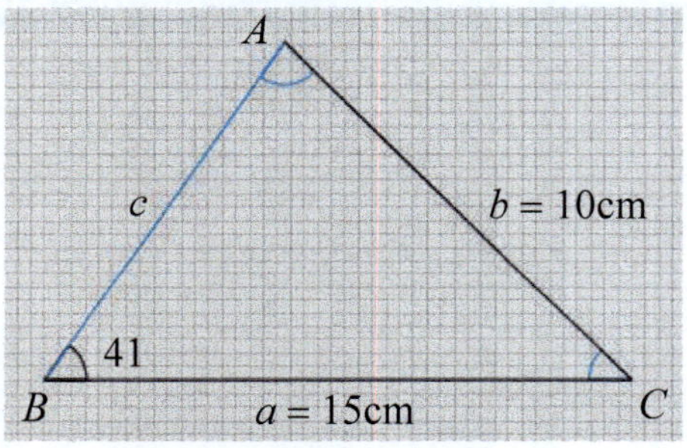

EXAM PRACTICE QUESTION 4.103 (ANSWER: $\hat{C} = 40^0$, $a = 35.2$ mm to 3 s.f., $c = 22.7$ mm to 3 s.f.)

In a triangle ABC, $\hat{A} = 95°$, $\hat{B} = 45°$ and $b = 25$ mm. Find the third angle and the other two sides.

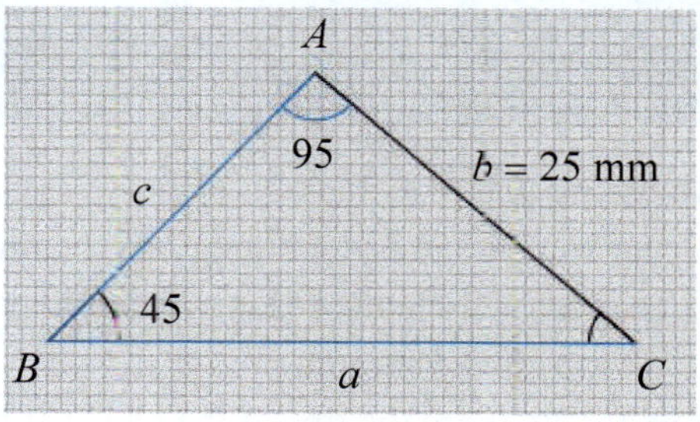

EXAM PRACTICE QUESTION 4.104 (ANSWERS: $a = 40.79$ m to 2 d.p., $k = 91.30$ m to 2 d.p., $h = 41.63$ m to 2 d.p., $\varphi = 24.07^0$, $\theta = 65.93^0$, $\omega = 114.07^0$, $x = 47.93^0$)

In the adjacent figure, where AB = 123 m, AD = 100 m $\measuredangle BAD = 18^0$ and $\measuredangle ACB = 90^0$

(i) Find the lengths a, k, h, and the angles φ, θ, ω, x

(ii) Check your results with Pythagoras' Theorem $(a+h)^2 + k^2 = (123)^2$

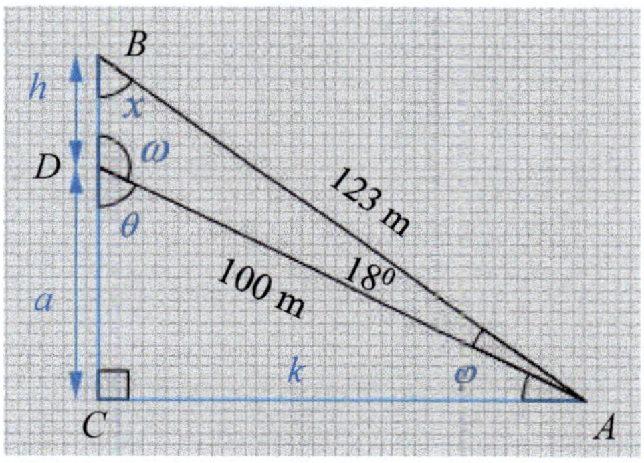

EXAM PRACTICE QUESTION 4.105 **(ANSWERS: (a) $x = 5.0$ cm, $A = 20.0$ cm^2, (b) $x = 2.0$ cm, $A = 0.0$ cm^2, (c) $x = 17.4$ cm, $A = 20$ cm^2, (d) $x = 15.6$ cm, $A = 34.6$ cm^2, (e) $x = 18.0$ cm, $A = 0.0$ cm^2, (f) $x = 17.4$ cm, $A = 20.0$ cm^2, (g) $x = 18.0$ cm, $A = 0.0$ cm^2, (h) $x = 17.4$ cm, $A = 20.0$ cm^2,)**

Assuming that in a clock:

i. the hours hand is 8 cm long and the minutes hand is 10 cm long.

ii. the entire circle (360°) is divided into 12 equal sectors representing the hours and that each sector is divided into 5 minutes

How far apart are the tips of the hands at the following times:

(a) 09:55, assuming that the hours hand is pointing to 10 and the minutes hand to 11

(b) 15:15, assuming that both hands are pointing to 3

(c) 17:00, where, the hours hand is pointing to 5, and the minutes hand is pointing to 12

(d) 16:00, where, the hours hand is pointing to 4, and the minutes hand is pointing to 12

(e) 18:00, where the hours hand is pointing to 6, and the minutes hand is pointing to 12

(f) 07:00, where the hours hand is pointing to 7, and the minutes hand is pointing to 12

(g) 21:15, assuming that the hours hand is pointing to 9, and the minutes hand is pointing to 3

(h) 10:25, assuming that the hours hand is pointing to 10, and the minutes hand is pointing to 5

Find the areas enclosed by the tips of the hands and the center of the clock.

Give all the answers with approximation to 1 d.p.

Topic 4.11 GEOMETRIC PROPERTIES OF A CIRCLE.

EXAM PRACTICE QUESTION 4.106

Draw a circle with radius 2.5 cm and name the different parts of the circle.

EXAM PRACTICE QUESTION 4.107 (ANSWER: 31.4 cm to 3 s.f.)

Work out the length of the arc of a semicircle of radius 10 cm, giving the answer to 3 s.f.

EXAM PRACTICE QUESTION 4.108 (ANSWER: 10.5 cm to 3 s.f.)

Workout the length of the arc of a sector subtending an angle of 30°, when $r = 20$ cm. Give the answer to three significant figures.

EXAM PRACTICE QUESTION 4.109 (ANSWERS: (a) 41.7 cm to 3 s.f., (b) 450 cm^2 (c) 707 cm^2 to 3 s.f., (d) 257 cm^2 to 3 s.f.)

The following figure shows a circle of radius 30 cm.

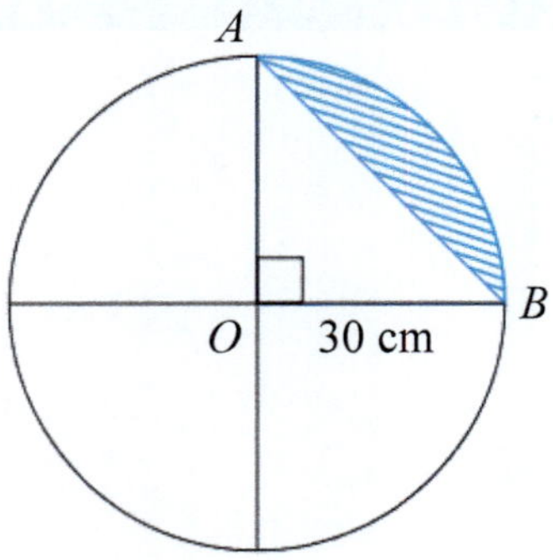

Find

(a) The length of the arc $\widehat{AB}$.
(b) The area of the triangle *OAB*.
(c) The area of the minor sector *OAB*.
(d) The area of the shaded segment.

EXAM PRACTICE QUESTION 4.110 (ANSWER: 2.62 mm to 3 s.f.)

In a circle with centre *O* and diameter 10 mm, find the length of an arc *AB* subtending an angle $\theta = 30°$.

EXAM PRACTICE QUESTION 4.111 (ANSWER: 28.7^0 to 3 s.f.)

In a circle of 50 cm radius, the length of an arc is 25 cm. Find the angle subtended by this arc in degrees.

EXAM PRACTICE QUESTION 4.112 **(ANSWER: 60^0, 15.7 mm and 300^0, 78.5 mm)**

A chord *AB* divides the circumference of a circle of diameter 30 mm in the ratio $\frac{s_1}{s_2} = \frac{1}{5}$, where s_1 and s_2 are the minor and major lengths of the corresponding arcs. Determine the angles subtended by the minor and major lengths of the arcs and the corresponding lengths of arcs.

EXAM PRACTICE QUESTION 4.113 **(ANSWER: 63.6 mm to 3 s.f.)**

A chord divides the circumference of a circle in the ratio $\frac{s_1}{s_2} = \frac{1}{3}$. Determine the length of the chord *AB* if the radius is 45 mm.

EXAM PRACTICE QUESTION 4.114 **(ANSWER: 77.40^0, 101 cm to 3 s.f.)**

A chord, *AB* of length 25 cm is drawn parallel to the diameter, *CD*, of a circle of radius 20 cm. Determine the angle at the centre $A\hat{O}B$ subtended by the chord and hence calculate the perimeter of *ABCDA*.

EXAM PRACTICE QUESTION 4.115 (ANSWER: 0.524 m^2 to 3 s.f.)

Calculate the area of the sector of a circle with $r = 1$ m, if $\theta = 60°$.

EXAM PRACTICE QUESTION 4.116 (ANSWER: 35.8^0 to 3 s.f.)

The area of a sector is 5 cm^2 and $r = 4$ cm. Find the angle θ.

EXAM PRACTICE QUESTION 4.117 (ANSWER: $\frac{A_{major\ segment}}{A_{minor\ segment}} = 265$ to the nearest whole number)

A chord divides the area of a circle into minor and major segments.
If $\theta = 30°$ is the angle of the minor segment, find the ratio of the major segment to that of the minor segment to the nearest whole number.

Topic 4.12 CIRCLE THEOREMS.

EXAM PRACTICE QUESTION 4.118 (ANSWER: $\widehat{OAN}=60^0$, $\widehat{COA}=150^0$)

In the following figure, workout $O\hat{A}N$ and $C\hat{O}A$, if $A\hat{O}N=30°$.

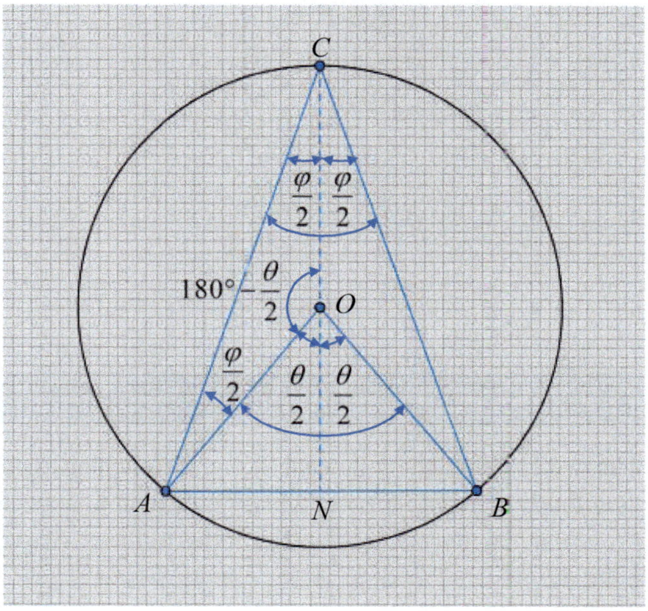

EXAM PRACTICE QUESTION 4.119 (ANSWER: $x=65^0, y=90^0$)

Workout the angle x and y in the following figures:

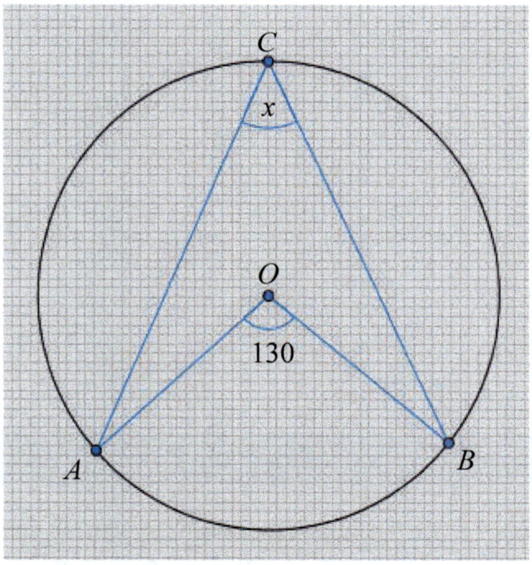

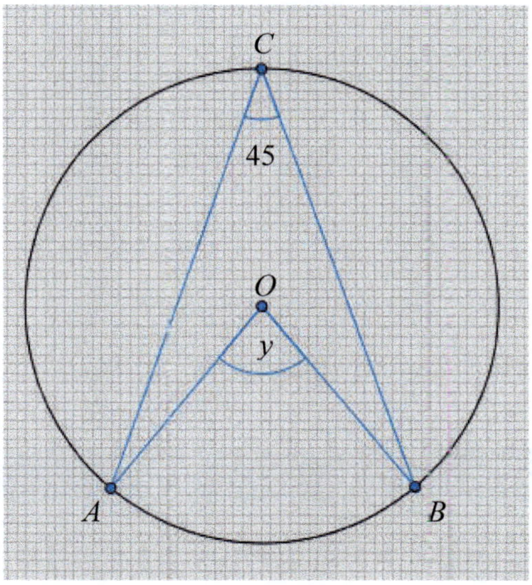

EXAM PRACTICE QUESTION 4.120 (ANSWER: $\theta = 135^0$, $\varphi = 112.5$)

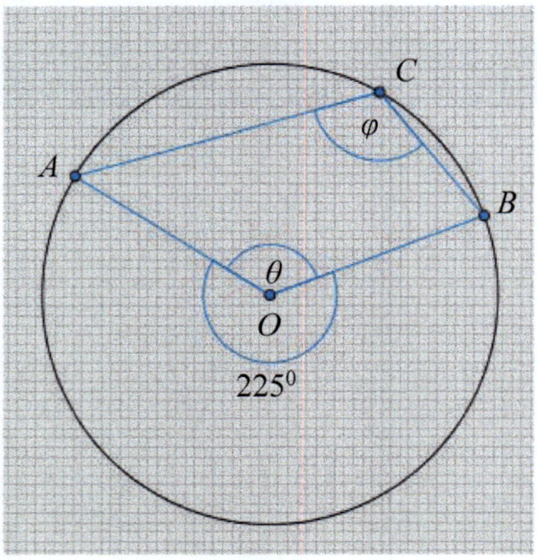

In the adjacent figure, find the angles θ and φ. Give reasons.

EXAM PRACTICE QUESTION 4.121

The two circles in the following figure are touching at point A. Line (f) is tangent to both circles at their touching point A. Also, line (ε) is tangent to both circles at their points B and D respectively. Prove that the line connecting the centres of the circles C_1 and C_2 is passing through their touching point A.

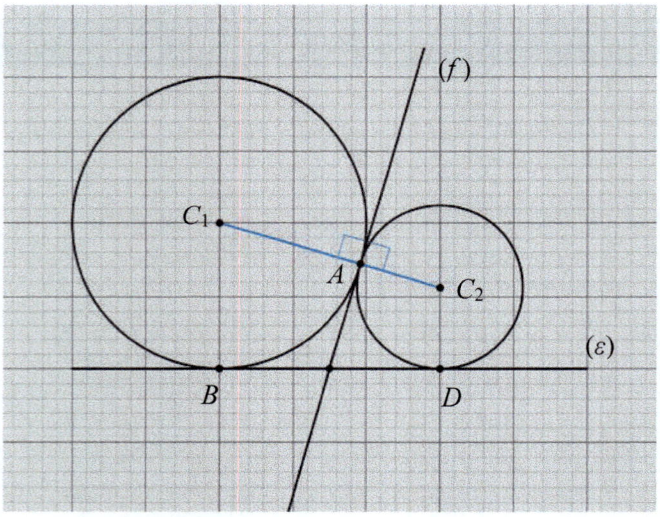

EXAM PRACTICE QUESTION 4.122 (ANSWER: 5 cm)

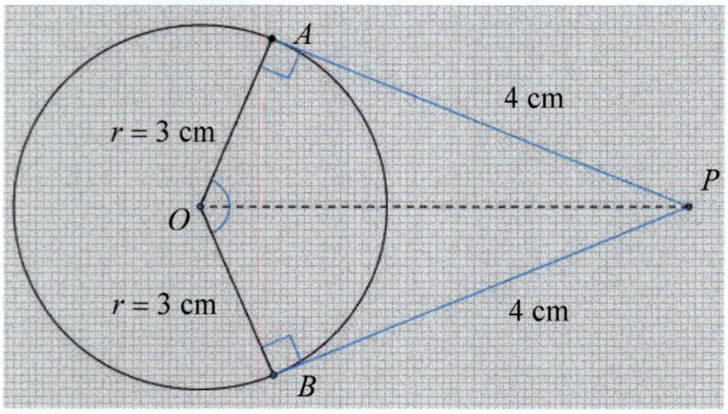

In the adjacent figure, the circle with centre O has radius r = 3 cm. PA and PB are tangents to the circle from P with lengths $PA = PB = 4$ cm. What is the distance between the point P and the centre of the circle?

EXAM PRACTICE QUESTION 4.123 (ANSWER: in the first figure, $\varphi = 45^0$, in the second figure, $\varphi = 37.5^0$, in the third figure, $\varphi = 30^0$)

In the figures below, *AB* is a diameter. Workout the measure of the angle marked φ.

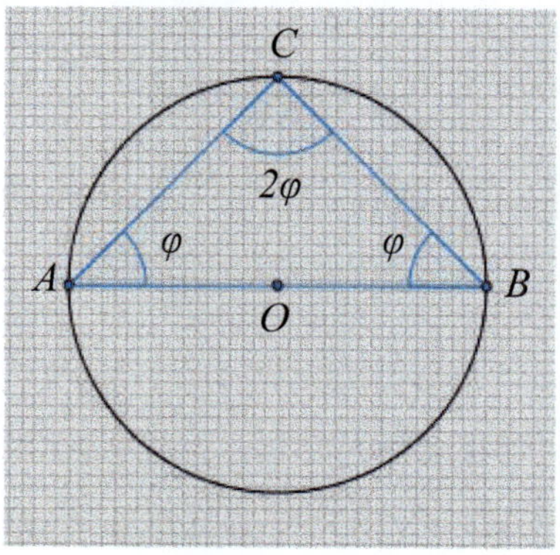

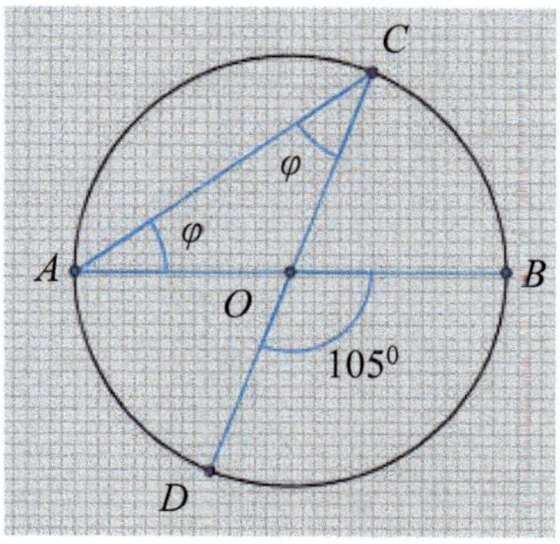

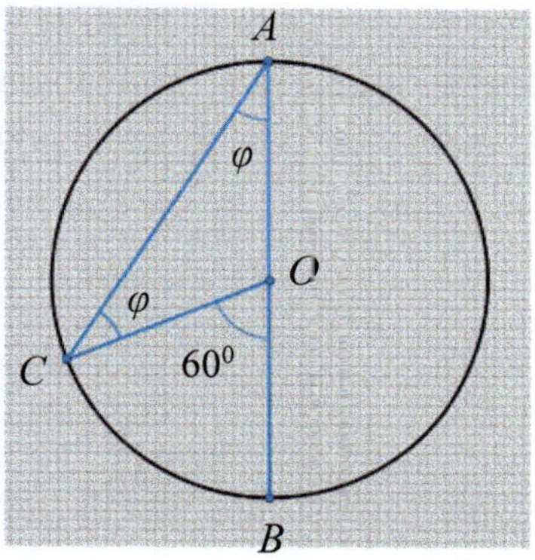

EXAM PRACTICE QUESTION 4.124 (ANSWER: 0.22 m^2)

In the following figure, line (ε) is tangent to the circle with centre *C* and radius $r = 0.5$ m, at point *A*. Given that the angle $\measuredangle BAF = 30^0$, find the area of the triangle *ABD*.

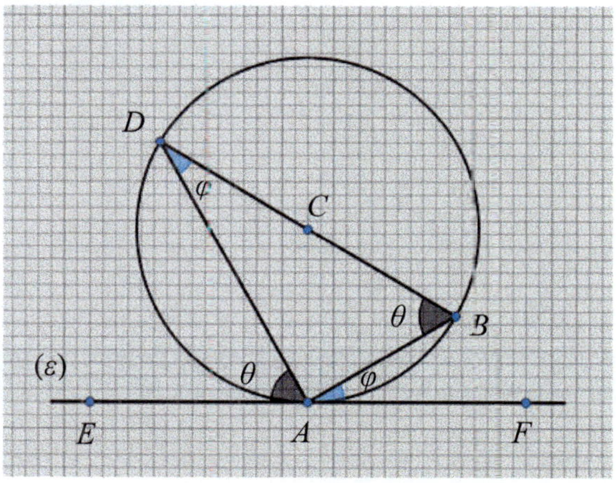

EXAM PRACTICE QUESTION 4.125

(ANSWER: (i) *ABED* is a rectangle, (ii) $AB = 2.40$ cm, $AD = 2.29$ cm, $A_{ABED} = 5.50$ cm^2 to 3 s.f.)

The two circles in the following figure are touching at point A. Line (f) is tangent to both circles at their touching point A. Also, line (ε) is tangent to both circles at their points B and D respectively. Tangent lines (f) and (ε) intersect at point M. On line (f), we pick a point E such that $AM = ME$.

(i) Determine the type of the quadrilateral *ABED*.

(ii) If $r_1 = 2$ cm, $r_2 = 1.3$ cm and $\measuredangle BC_1A = \theta = 73.8^0$, find the lengths of the sides of the quadrilateral *ABED* and its surface area.

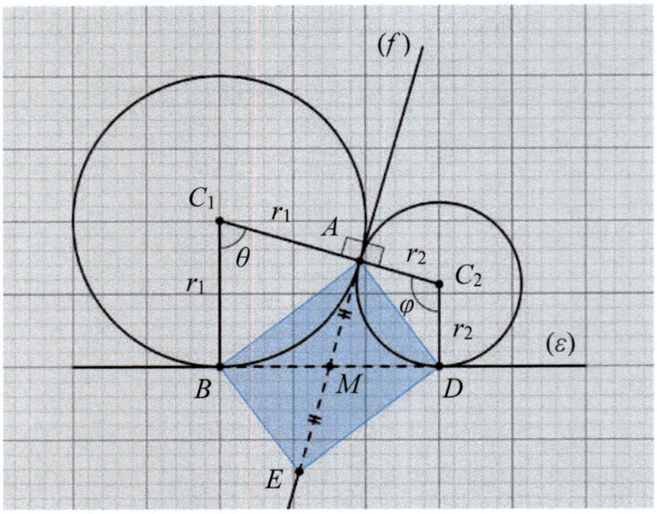

EXAM PRACTICE QUESTION 4.126

(ANSWER: 60^0)

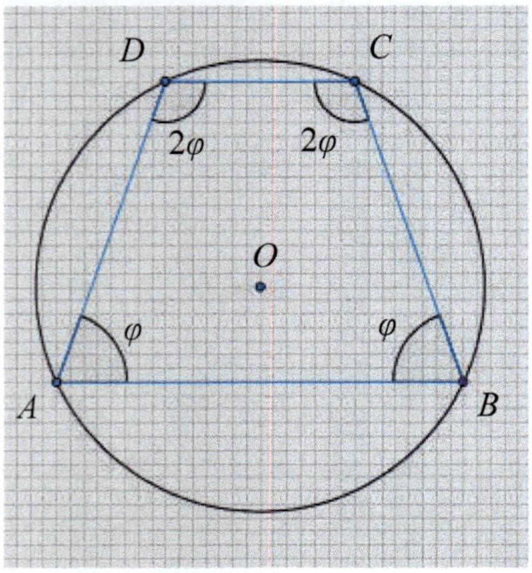

In the adjacent figure, workout the angle φ.

EXAM PRACTICE QUESTION 4.127 (ANSWER: $\varphi = 60^0, \theta = 100^0$)

In the adjacent figure, workout the angles φ and θ.

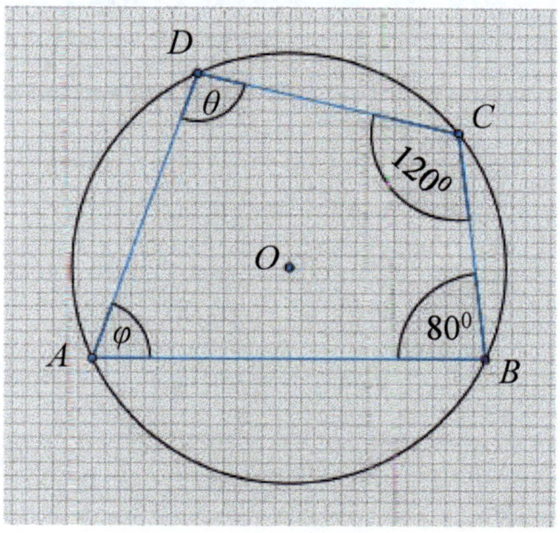

EXAM PRACTICE QUESTION 4.128

In the adjacent figure prove that
$\measuredangle EAD = \measuredangle DCB = \varphi$ and $\measuredangle FBA = \measuredangle CDA = \theta$.

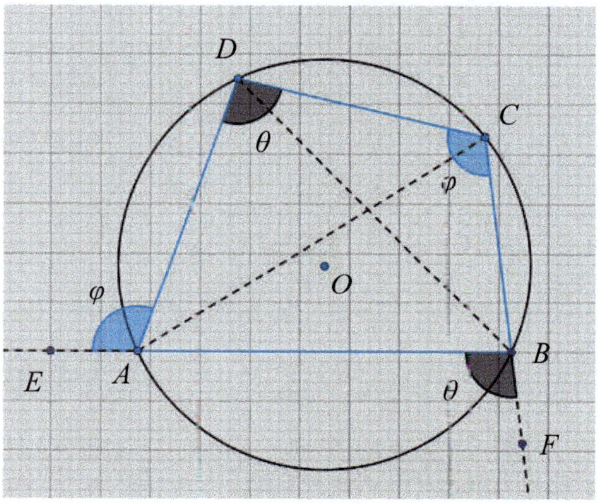

Topic 4.13 SURFACE AREAS AND VOLUMES OF SPHERES, CONES CYLINDERS AND CUBOIDS.

In the figures below write the formulae for their surface areas and their volumes.

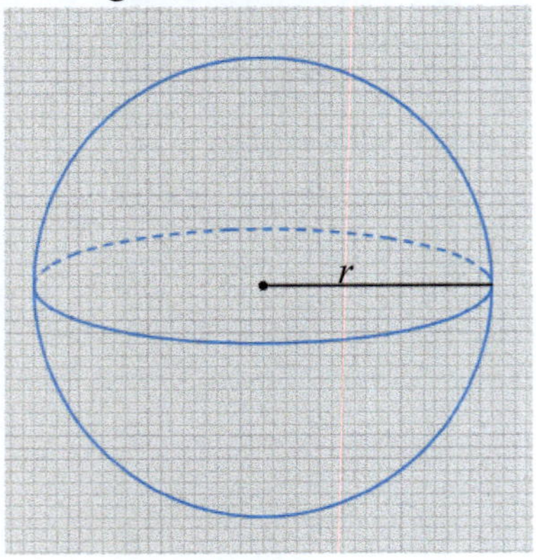

Sphere

Surface area of the *sphere*:

Volume of the sphere:

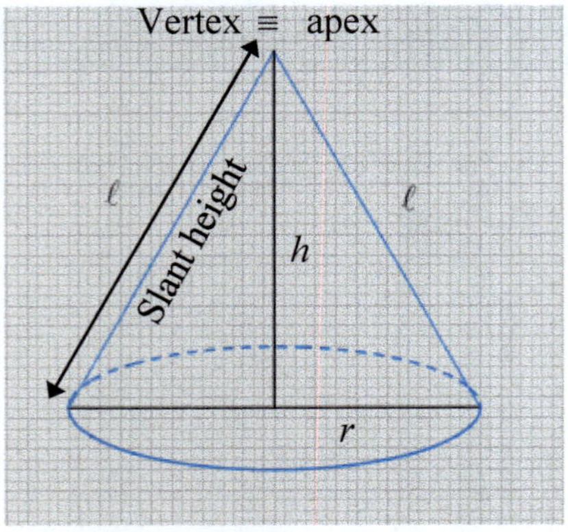

Right Cyclical Cone, or Cone

Surface area of the *cone*:

Volume of the cone:

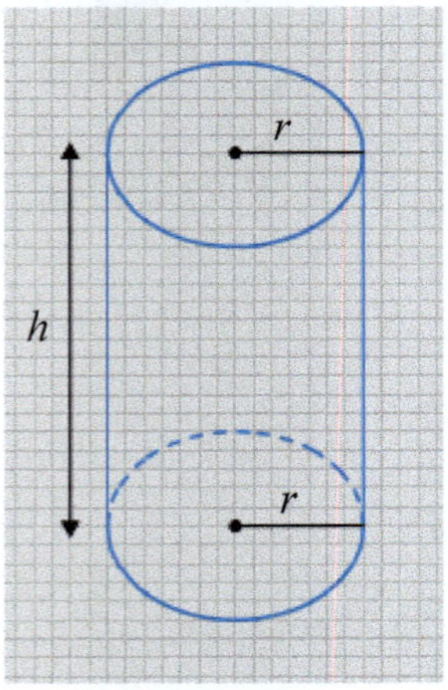

Cylinder

Surface area of the cylinder:

Volume of the cylinder:

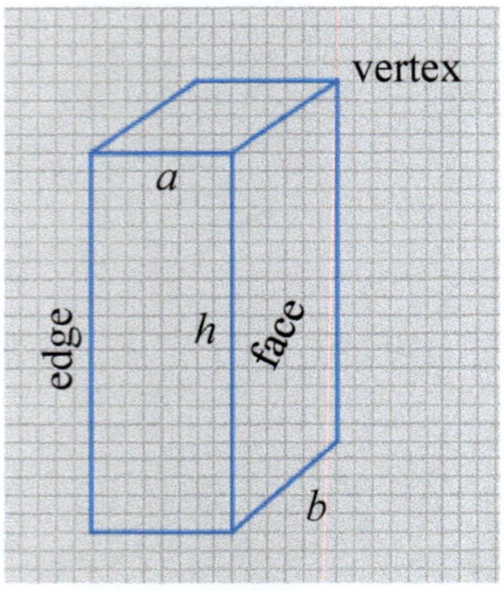

Cuboid

Surface area of the cuboid:

Volume of the cuboid:

EXAM PRACTICE QUESTION 4.129 (ANSWER: $S = 220$ cm^2, $V = 200$ cm^3)

Calculate the surface area of the cuboid shown in the adjacent figure and its volume.

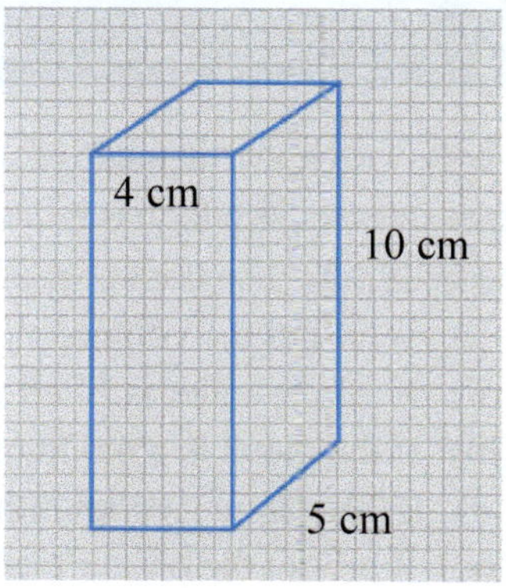

EXAM PRACTICE QUESTION 4.130 (ANSWERS: (i) $S = 700\pi$ cm^2, (ii) $V = 8\,000\pi/3$ cm^3)

A hemisphere of radius r = 10 cm fits exactly on top of a cylinder of height h = 20 cm, as shown in the adjacent figure.

(a) Find the surface area of this shape.
(b) Workout the volume. Give your answers in terms of π.

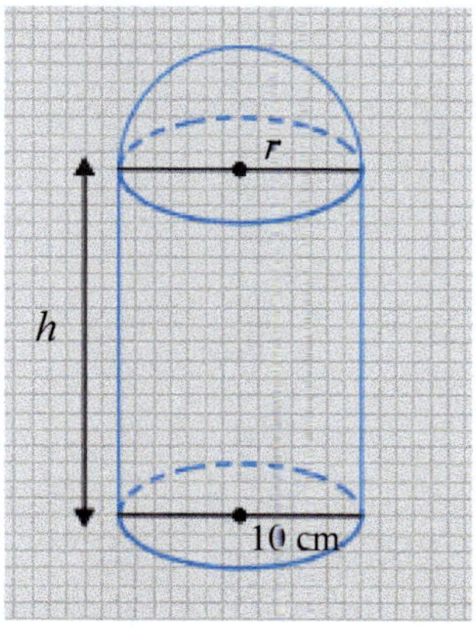

Exam Practice Question 4.131 **(Answers: (i) V_{metal} = 53.0 cm^3 to 3 s.f., (ii) V_{water} = 88.4 cm^3 to 3 s.f., (iii) W_{total} = 2.3 newton to 1 d.p., (iv) P_{total} = 813.5 N/m^2 to 1 d.p)**

A hollow cylinder is made of a metal sheet of thickness 0.5 cm.
The external cylinder has radius $r_1 = 3$ cm and height $h_1 = 5$ cm and the internal cylinder has radius r_2 cm and height h_2. The thickness of the base of the cylinder is the same with the thickness of its wall.

(i) Workout the volume of the metal of the cylinder with an open top.
(ii) If this cylinder is filled with water, what is the volume of the water it can contain.
(iii) If the metal is aluminum with density 2 700 kg/m^3 and the water is distilled with density 1 000 kg/m^3, workout the weight of the filled with water cylinder. Take $g = 9.81$ m/s^2
(iv) Workout the pressure applied on the base of the filled with water cylinder.

EXAM PRACTICE QUESTION 4.132

A cone is produced from a sector of a circle of angle $\theta°$.
Prove that the curved surface area of a cone is given by $S = \pi r\ell$ where r is the radius of the base and ℓ is the slant height.

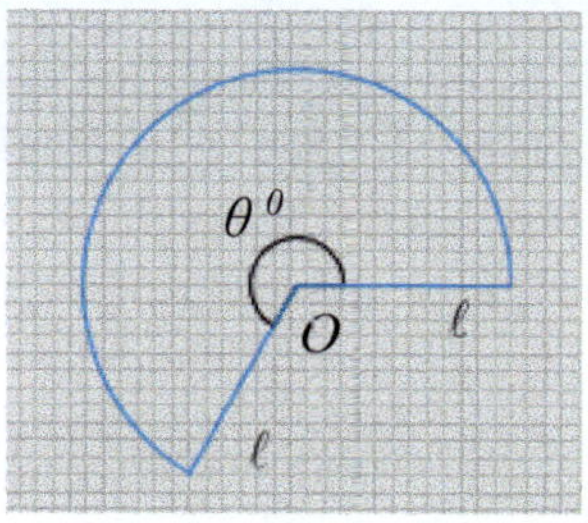

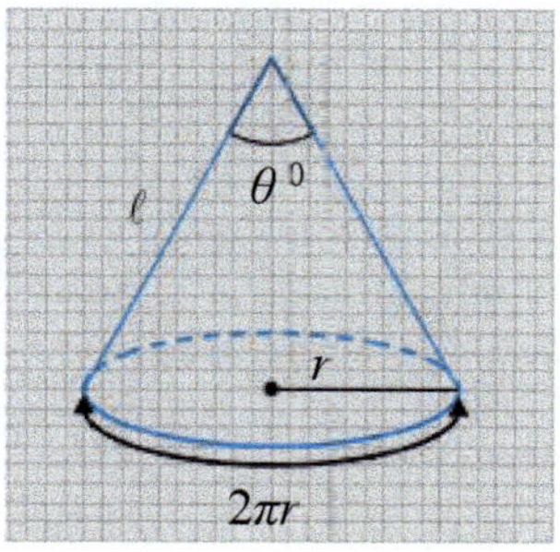

EXAM PRACTICE QUESTION 4.133

(ANSWERS: (a) $r = 3.75$ cm, (b) $l = R = 5$ cm, (c) $V = 48.7\ cm^3$ to 3 s.f., (d) $S = 58.9\ cm^2$ to 3 s.f.)

The sector of a circle is shown in the adjacent figure. This shape is made into a cone.

(a) Find the radius of the base of the cone.
(b) Write down the length of the slant.
(c) Determine the volume of the cone.
(d) Workout the surface area of the curved part.

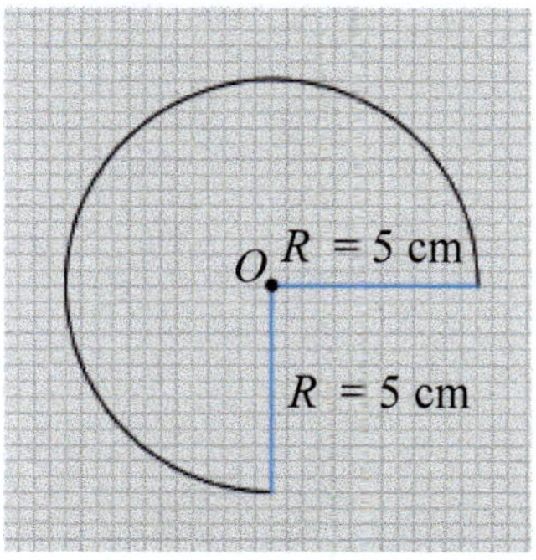

EXAM PRACTICE QUESTION 4.134 (ANSWERS: (a) r_2:r_1 = 1:2, (b) $S_1 = 400\pi\left(1+\sqrt{2}\right)$, $S_2 = 100\pi\left(1+\sqrt{2}\right)$, (c) $S_{Frustum\ Cone} = 100\pi\left(5+3\sqrt{2}\right)$, (d) $V_1 = \frac{8\ 000\pi}{3}$, $V_2 = \frac{1\ 000\pi}{3}$, (e) $V_{Frustum\ Cone} = \frac{7\ 000\pi}{3}$)

A small cone of height h_2 = 10 cm and radius r_2 = 10 cm is truncated from a large cone of height h_1 = 20 cm and radius r_1 = 20 cm, leaving a frustum cone as shown in the following figure.

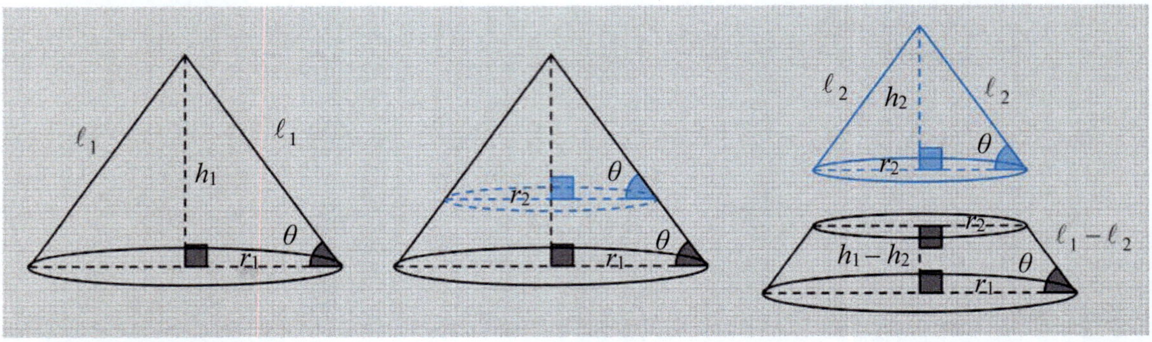

FIG. 4.13.13

Determine:

(a) The ratio of the radii of the bases of the cones.
(b) The surface areas of the large cone of height h_1 and the small cone of height h_2.
(c) The surface area of the frustum cone.
(d) The volumes of the large and the small cones.
(e) The volume of the frustum cone.
Give your answers in terms of π.

Topic 4.14 PYTHAGORAS' THEOREM IN THREE DIMENSIONS.

EXAM PRACTICE QUESTION 4.135 (ANSWERS: (i) $AC = 14.1$ cm, $AG = 17.3$ cm to 3 s.f., (ii) $AC = 7.81$ cm to 3 s.f., $CE = 16.9$ cm to 3 s.f.)

(i) For Fig. 4.14.1, apply Pythagoras' theorem to find the diagonal AC. Hence apply the theorem again to find AG.

(ii) For Fig. 4.14.2 apply Pythagoras' theorem to find the diagonal AC hence apply the theorem again to find CE.

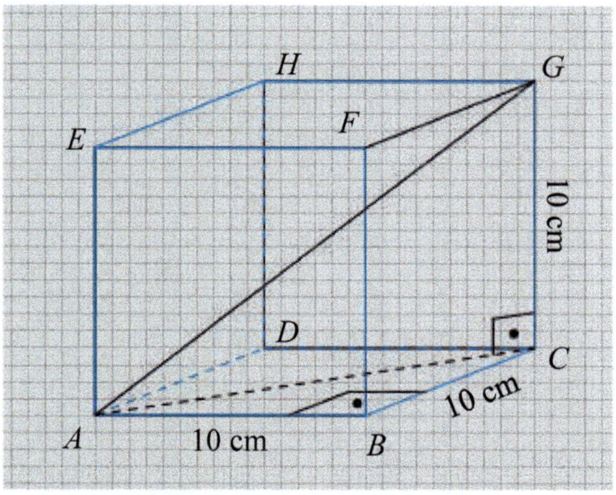

FIG. 4.14.1 A *cube* of side 10 cm

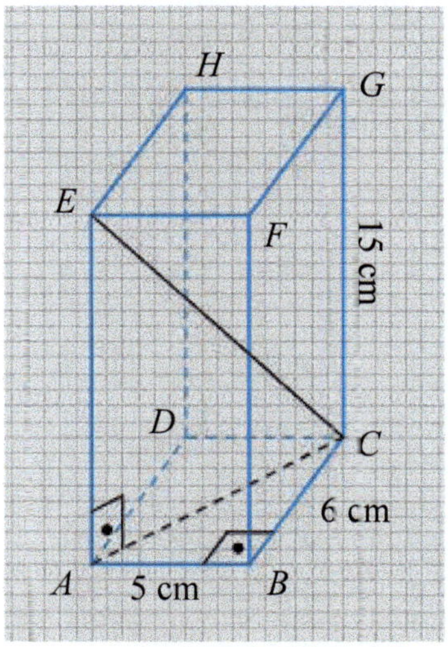

FIG. 4.14.2 A *cuboid* or *prism* of Base 5 cm × 6 cm & height 15 cm

EXAM PRACTICE QUESTION 4.136 (ANSWER: $BV = 11.7$ cm to 3 s.f., $AC = 17.0$ cm to 3 s.f.)

For the square base pyramid in the adjacent figure, find the size of the edge, $BV = AV = DV = CV$. Also, Calculate the diagonal AC.

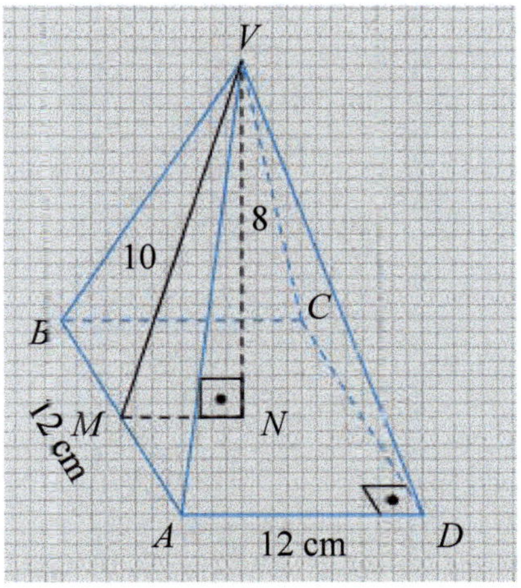

EXAM PRACTICE QUESTION 4.137 (ANSWER: $l = 11.7$ cm to 3 s.f.)

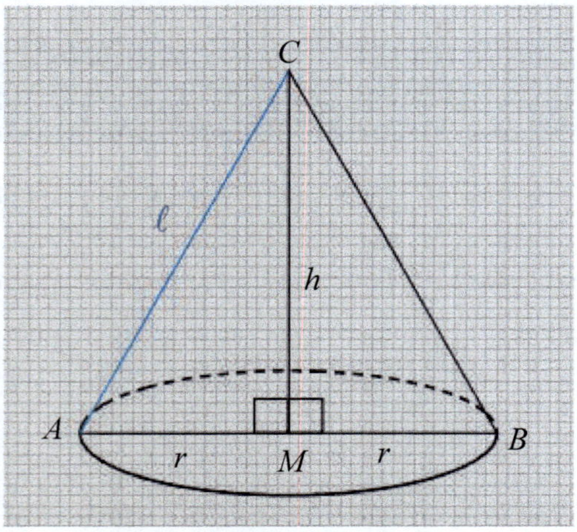

Workout the slant height ℓ in the right circular cone of the adjacent figure, where $r = 5$ cm and $h = 10$ cm.

Topic 4.15 SIMILARITY RULES FOR 2-DIMENSIONAL SHAPES AND 3-DIMENSIONAL SOLIDS. AREAS & VOLUMES.

Similarity Rules for 2-Dimensional Shapes

Similar triangles

Let the similarity ratio be k , that is:
$k = \ell_2 : \ell_1 = h_2 : h_1$

Calculate the ratio of the areas of the two similar triangles:

$\frac{A_2}{A_1} =$

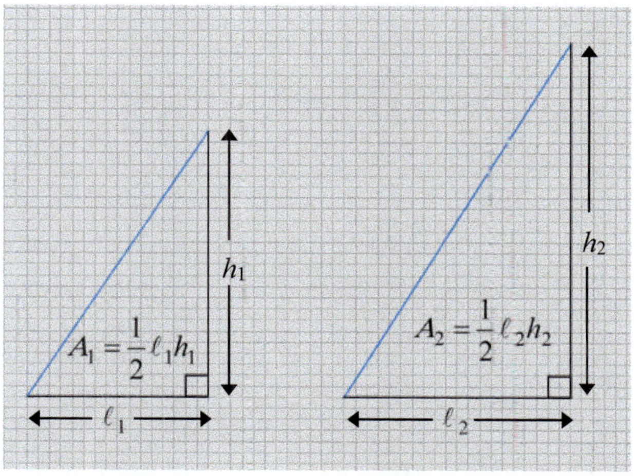

Similar Squares

Let the similarity ratio be k, that is:
$k = \ell_2 : \ell_1$

Calculate the ratio of the areas of the two similar squares:

$\frac{A_2}{A_1} =$

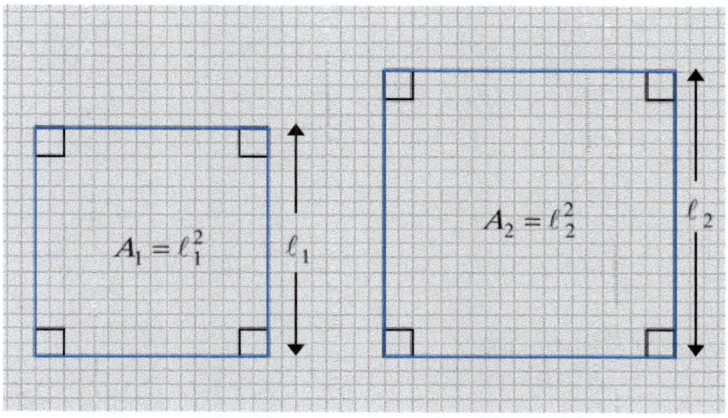

Similar rectangles

Let the similarity ratio be k, that is:
$k = \ell_2 : \ell_1 = w_2 : w_1$

Calculate the ratio of the areas of the two similar rectangles:

$\frac{A_2}{A_1} =$

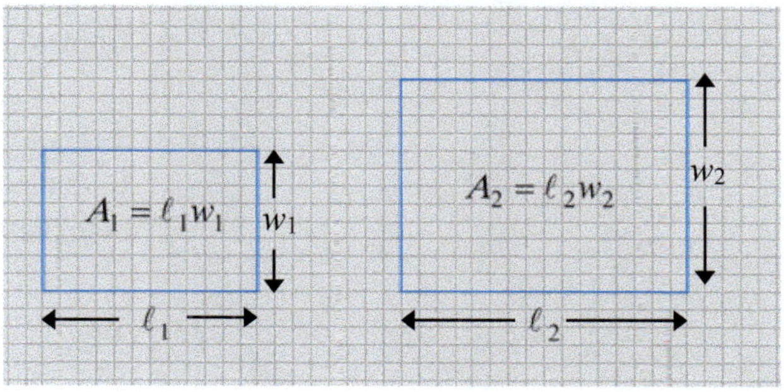

Similar Circles

Let the similarity ratio be k, that is:
$k = r_2 : r_1$

Calculate the ratio of the areas of the two similar circles:

$\frac{A_2}{A_1} =$

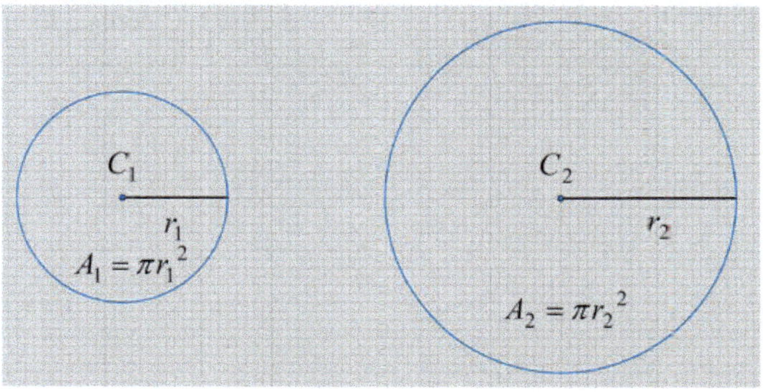

Similarity Rules for 3-Dimensional Solids

In the 3D shapes we consider both, areas and volumes.

Similar Spheres

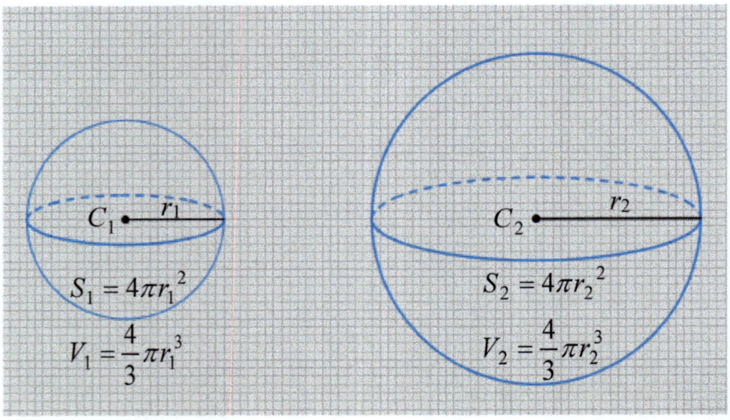

Let the similarity ratio be *k*, that is:
$k = r_2 : r_1$

Surface areas
Calculate the ratio of the surface areas of the two similar spheres:

$$\frac{S_2}{S_1} =$$

Volumes
Calculate the ratio of the volumes of the two similar spheres:

$$\frac{V_2}{V_1} =$$

Similar Cylinders

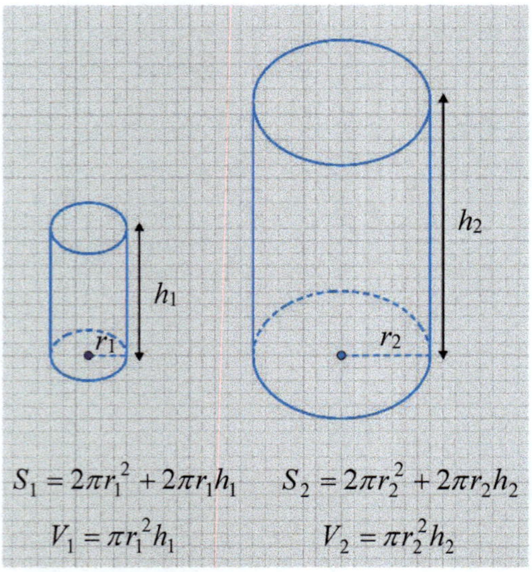

Let the similarity ratio be *k*, that is: $k = r_2 : r_1 = h_2 : h_1$,
from where $r_2 = kr_1$ and $h_2 = kh_1$

Surface areas
Calculate the ratio of the surface areas of the two similar cylinders:

$$\frac{S_2}{S_1} =$$

Volumes
Calculate the ratio of the volumes of the two similar cylinders:

$$\frac{V_2}{V_1} =$$

Similar right circular cones.

Let the similarity ratio be k, that is:

$k = a_2 : a_1 = h_2 : h_1$, from where

$a_2 = ka_1$ and $h_2 = kh_1$

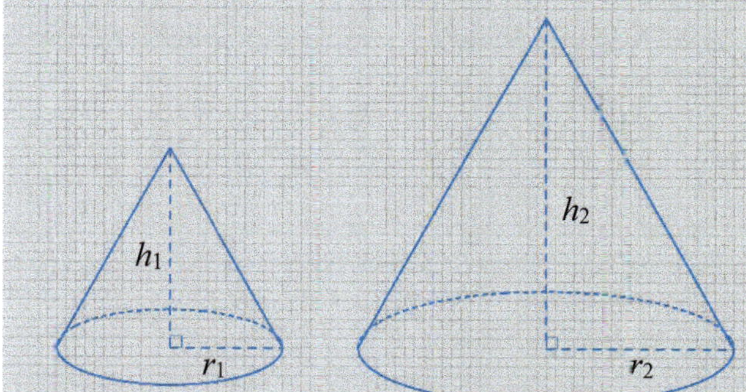

$$S_1 = \pi r_1^2 + \pi r_1 \ell_1 = \pi r_1^2 + \pi r_1\sqrt{r_1^2 + h_1^2}$$

$$S_2 = \pi r_2^2 + \pi r_2 \ell_2 = \pi r_2^2 + \pi r_2\sqrt{r_2^2 + h_2^2}$$

$$V_1 = \frac{1}{3}\pi r_1^2 h_1 \qquad V_2 = \frac{1}{3}\pi r_2^2 h_2$$

Surface areas

Calculate the ratio of the surface areas of the two similar cones:

$\frac{S_2}{S_1} =$

Volumes

Calculate the ratio of the volumes of the two similar cones:

$\frac{V_2}{V_1} =$

Similar Square Pyramids

Let the similarity ratio be k, that is:

$k = a_2 : a_1 = h_2 : h_1$, from where

$a_2 = ka_1$ and $h_2 = kh_1$

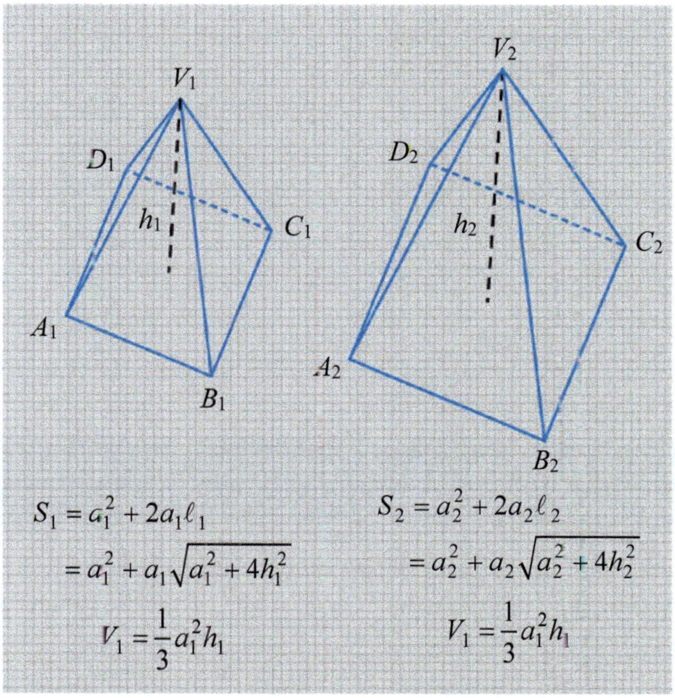

$$S_1 = a_1^2 + 2a_1\ell_1 = a_1^2 + a_1\sqrt{a_1^2 + 4h_1^2}$$

$$S_2 = a_2^2 + 2a_2\ell_2 = a_2^2 + a_2\sqrt{a_2^2 + 4h_2^2}$$

$$V_1 = \frac{1}{3}a_1^2 h_1 \qquad V_1 = \frac{1}{3}a_1^2 h_1$$

Surface areas

Calculate the ratio of the surface areas of the two similar square pyramids:

$\frac{S_2}{S_1} =$

Volumes

Calculate the ratio of the volumes of the two similar square pyramids:

$\frac{V_2}{V_1} =$

EXAM PRACTICE QUESTION 4.138 (ANSWER: $\frac{\sqrt{5}}{1}$)

The increase of the area of a picture after an enlargement is 400%. Determine the scale factor of the enlargement.

EXAM PRACTICE QUESTION 4.139 (ANSWERS: (a) $\frac{3}{2}$, (b) $\frac{3}{2}$, (c) YES, (d) $\frac{9}{4}$, (e) $\frac{27}{8}$)

Two right cylindrical cans of radii 10 cm and 15 cm have corresponding heights of 50 cm and 75 cm. Determine:

(a) The ratio of the radii

(b) The ratio of the heights

(c) Are the cans two similar solids?

(d) The ratio of the surface areas

(e) The ratio of the volumes.

EXAM PRACTICE QUESTION 4.140 (ANSWERS: (a) $\frac{5}{1}$, (b) $\frac{25}{1}$)

The ratio of the volumes of two similar spheres is 125 : 1. Find:

(a) The ratio of their radii and (b) The ratio of their surface areas.

EXAM PRACTICE QUESTION 4.141 (ANSWERS: (a) $\frac{3}{2}$, (b) $\frac{3}{2}$)

Two similar flower pots have volumes in the ratio 27 : 8. Find:

(a) The ratio of their heights

(b) The ratio of their radii.

EXAM PRACTICE QUESTION 4.142 (ANSWER: $d_2 = 12$ cm)

Two similar bowls have volumes in the ratio 8 : 1. If the diameter of the small bowl is 6 cm, find the diameter of the large bowl.

Topic 4.16 THREE DIMENSIONAL OBJECTS. CONSTRUCTION OF PLANS AND ELEVATIONS OF 3D OBJECTS.

EXAM PRACTICE QUESTION 4.143

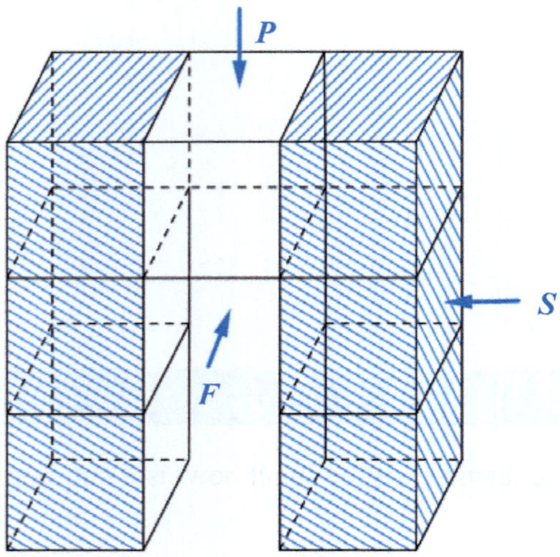

Show the plan and elevation views (front elevation, side elevation) for the 3D – object in the adjacent figure.

EXAM PRACTICE QUESTION 4.144

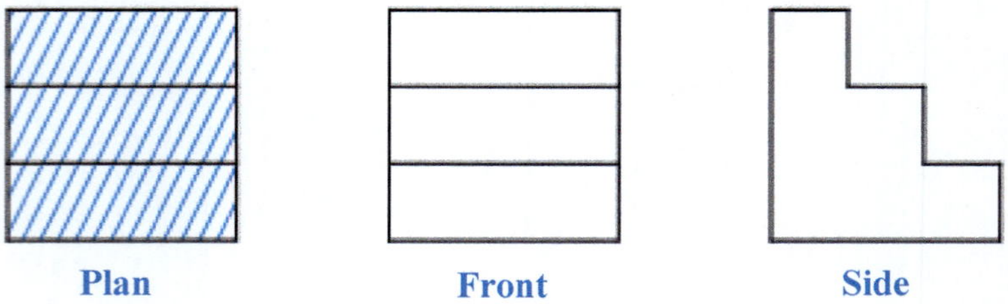

Draw the 3 – D object that gives the views as shown in the adjacent figure.

Topic 4.17 VECTORS.

EXAM PRACTICE QUESTION 4.145 (ANSWER: $\begin{pmatrix}6\\8\end{pmatrix}$)

Find the sum of the vectors $\begin{pmatrix}2\\3\end{pmatrix}$ and $\begin{pmatrix}4\\5\end{pmatrix}$.

EXAM PRACTICE QUESTION 4.146 (ANSWER: $\begin{pmatrix}0\\6\end{pmatrix}$)

Find the addition of the vectors $\begin{pmatrix}-3\\4\end{pmatrix}$ and $\begin{pmatrix}3\\2\end{pmatrix}$.

EXAM PRACTICE QUESTION 4.147 (ANSWERS: (i) $\begin{pmatrix}8\\12\end{pmatrix}$, (ii) $\begin{pmatrix}\frac{2}{3}\\1\end{pmatrix}$, (iii) $\begin{pmatrix}-6\\-9\end{pmatrix}$)

If $\mathbf{a} = \begin{pmatrix}2\\3\end{pmatrix}$, find (i) $4\mathbf{a}$, (ii) $\frac{1}{3}\mathbf{a}$, (iii) $-3\mathbf{a}$

EXAM PRACTICE QUESTION 4.148 (ANSWERS: (i) 25, (ii) 35)

If the vector **a** has magnitude 5, find the magnitude (i) of $5\mathbf{a}$ and (ii) of $-7\mathbf{a}$.

EXAM PRACTICE QUESTION 4.149 (ANSWERS: (i) 5, (ii) 13, (iii) 10)

Find the magnitude of each one of the following vectors:

(i) $\begin{pmatrix} 3 \\ 4 \end{pmatrix}$ (ii) $\begin{pmatrix} 5 \\ 12 \end{pmatrix}$ (iii) $\begin{pmatrix} -6 \\ -8 \end{pmatrix}$

EXAM PRACTICE QUESTION 4.150 (ANSWERS: (a)40 km, (ii) O)

An athlete runs 10 km in the north direction and then 10 km in the east direction, 10 km in the south direction and 10 km in the west direction. Find (a) the total distance covered (b) the total displacement.

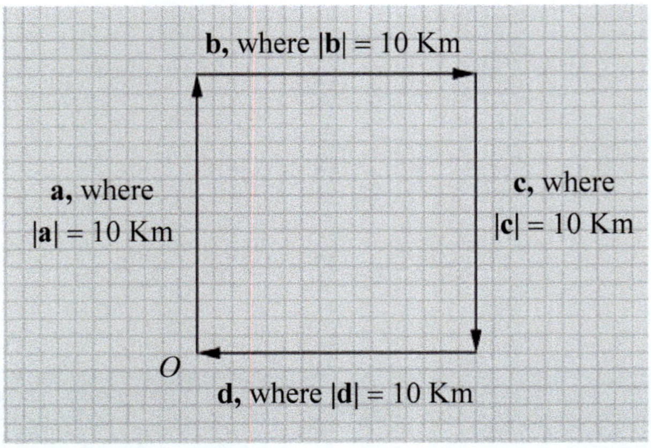

EXAM PRACTICE QUESTION 4.151 (ANSWER: $\overrightarrow{AB} = \overrightarrow{AC} + \overrightarrow{CB}$, $\overrightarrow{A_1B_1} = \overrightarrow{A_1C_1} + \overrightarrow{C_1B_1}$, $\left|\overrightarrow{AB}\right| = \sqrt{13}$, $\left|\overrightarrow{AC}\right| = 2$, $\left|\overrightarrow{CB}\right| = 3$, $\left|\overrightarrow{A_1B_1}\right| = \sqrt{13}$, $\left|\overrightarrow{A_1C_1}\right| = 3$, $\left|\overrightarrow{C_1B_1}\right| = 2$)

Describe the movement from A to B and from A_1 to B_1 in the following figure, using the vector notation. Then find the magnitude (or modulus) of each vector.

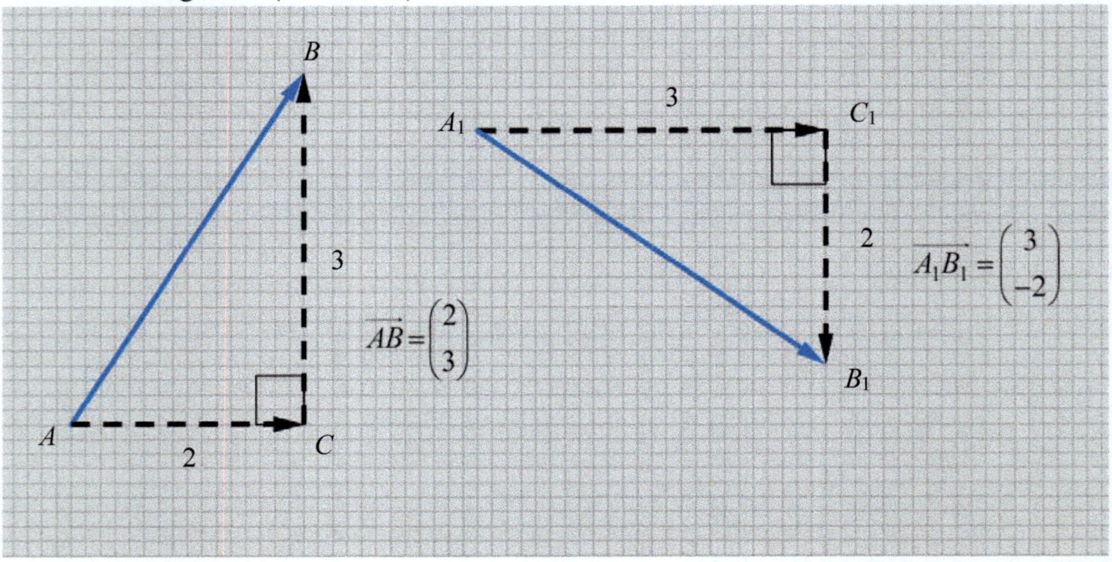

EXAM PRACTICE QUESTION 4.152 (ANSWER: $\overrightarrow{OA}=\begin{pmatrix}-1\\-3\end{pmatrix}$, $\overrightarrow{OB}=\begin{pmatrix}2\\4\end{pmatrix}$, $\overrightarrow{OC}=\begin{pmatrix}-3\\5\end{pmatrix}$, $\overrightarrow{OD}=\begin{pmatrix}-4\\3\end{pmatrix}$)

The points *A, B, C, D* have the following coordinates: $A(-1, -3)$, $B(2, 4)$, $C(-3, 5)$ and $D(-4, 3)$. Plot these points and write down the position vectors relative to the origin of *A, B, C* and *D*.

EXAM PRACTICE QUESTION 4.153 (ANSWERS: (a) $\overrightarrow{OA}=\begin{pmatrix}4\\6\end{pmatrix}$, $\overrightarrow{OB}=\begin{pmatrix}8\\2\end{pmatrix}$, $\overrightarrow{OC}=\begin{pmatrix}11\\9\end{pmatrix}$, (b) $\overrightarrow{AB}=\begin{pmatrix}4\\-4\end{pmatrix}$, $\overrightarrow{CB}=\begin{pmatrix}-3\\-7\end{pmatrix}$, $\overrightarrow{AC}=\begin{pmatrix}7\\3\end{pmatrix}$)

Consider the points *A*, *B* and *C* with coordinates $A(4, 6)$, $B(8, 2)$ and $C(11, 9)$.

(a) Draw *A*, *B* and *C* and determine the position and the free vectors between these points.

(b) Write in column vector notation the vectors $\overrightarrow{AB}, \overrightarrow{CB}, \overrightarrow{AC}$.

EXAM PRACTICE QUESTION 4.154

(ANSWER: $\overrightarrow{AB}=\begin{pmatrix}2\\0\end{pmatrix}$, $\overrightarrow{BC}=\begin{pmatrix}1\\-5\end{pmatrix}$, $\overrightarrow{AC}=\begin{pmatrix}3\\-5\end{pmatrix}$)

The position vectors $\overrightarrow{OA}, \overrightarrow{OB}$ and $\overrightarrow{OC}$ are given as $\overrightarrow{OA}=\begin{pmatrix}3\\4\end{pmatrix}$, $\overrightarrow{OB}=\begin{pmatrix}5\\4\end{pmatrix}$ and $\overrightarrow{OC}=\begin{pmatrix}6\\-1\end{pmatrix}$.

Plot these points and find the free vectors $\overrightarrow{AB}$, $\overrightarrow{BC}$ and $\overrightarrow{AC}$.

EXAM PRACTICE QUESTION 4.155

(ANSWERS: (i) $\overrightarrow{PQ}=\begin{pmatrix}-5\\0\end{pmatrix}$, (ii) $|\overrightarrow{PQ}|=5$)

Plot the vectors $\overrightarrow{OP}=\begin{pmatrix}7\\5\end{pmatrix}$ and $\overrightarrow{OQ}=\begin{pmatrix}2\\5\end{pmatrix}$

(i) Find the vector $\overrightarrow{PQ}$ (ii) Find the magnitude of $\overrightarrow{PQ}$

EXAM PRACTICE QUESTION 4.156

(ANSWERS: (i) $\overrightarrow{AB} = \begin{pmatrix} 0 \\ 4 \end{pmatrix}$, (ii) $|\overrightarrow{AB}| = 4$)

Plot the vectors $\overrightarrow{OA} = \begin{pmatrix} 2 \\ 5 \end{pmatrix}$ and $\overrightarrow{OB} = \begin{pmatrix} 2 \\ 9 \end{pmatrix}$

(i) Write $\overrightarrow{AB}$ as a column vector, (ii) Find the modulus of $\overrightarrow{AB}$

EXAM PRACTICE QUESTION 4.157

(ANSWERS: (i) $\overrightarrow{BC} = $ b, (ii) $\overrightarrow{CD} = -$ a, (iii) $\overrightarrow{AC} = $ a+b)

ABCD is a parallelogram. The vector from *A* to *B* is **a**. The vector from *A* to *D* is **b**.

State: (i) $\overrightarrow{BC}$, (ii) $\overrightarrow{CD}$, (iii) $\overrightarrow{AC}$

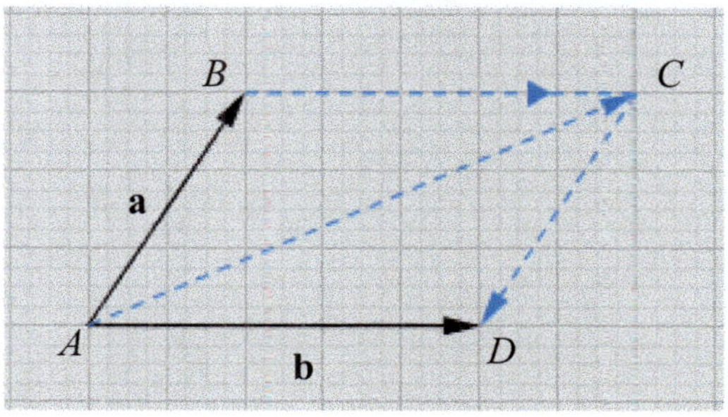

EXAM PRACTICE QUESTION 4.158

(ANSWERS: $\overrightarrow{BA} = 2$(a-b), $\overrightarrow{NM} = $ a-b)

In the following figure *OMA* and *ONB* are straight lines. *MN* and *AB* are parallel lines.

$\overrightarrow{OA} = 2\mathbf{a}$ and $\overrightarrow{OB} = 2\mathbf{b}$. Express $\overrightarrow{BA}$ and $\overrightarrow{NM}$ in terms of **a** and **b**. *M* and *N* are the midpoints of *OA* and *OB* respectively.

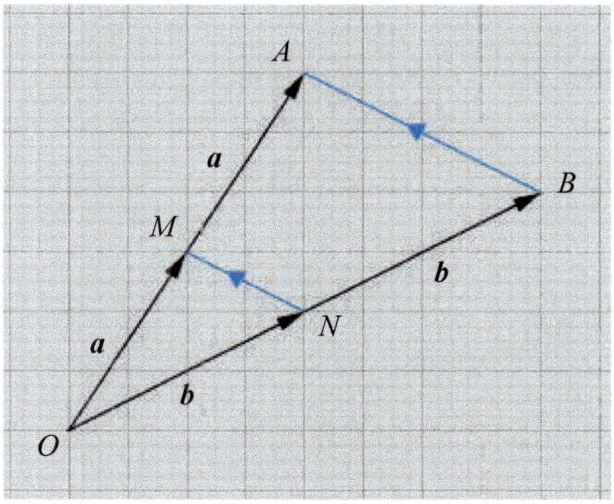

EXAM PRACTICE QUESTION 4.159 (ANSWERS: (i) $\overrightarrow{AB}=\begin{pmatrix}-4\\3\end{pmatrix}$, (ii) $|\overrightarrow{AB}|=5$)

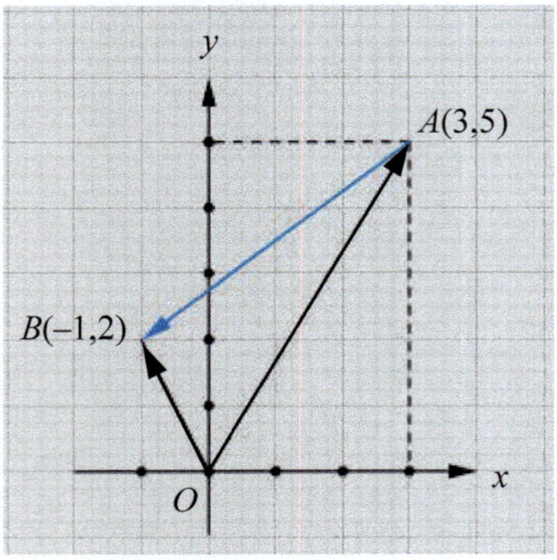

A is the point (3, 5) and B is the point (−1, 2).

(i) Write $\overrightarrow{AB}$ as a column vector.

(ii) Find the length of the vector $\overrightarrow{AB}$

EXAM PRACTICE QUESTION 4.160 (ANSWERS: (i) $\overrightarrow{AB}=$ b - a, (ii) $\overrightarrow{MN}=\frac{1}{2}$(b - a))

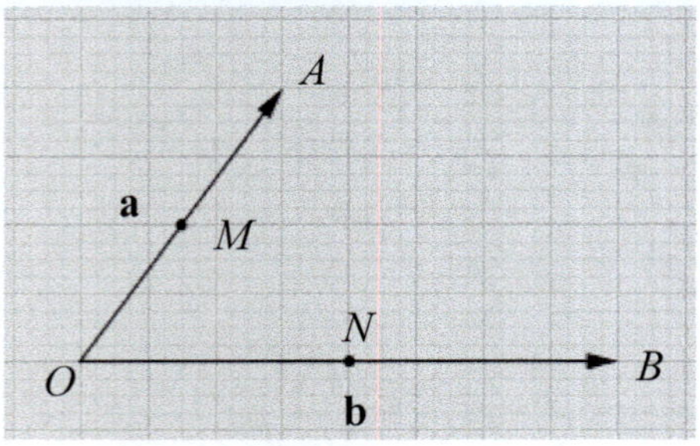

The adjacent figure shows the position vectors $\overrightarrow{OA}$ and $\overrightarrow{OB}$. Find:

(i) The free vector $\overrightarrow{AB}$.

(ii) The free vector $\overrightarrow{MN}$, where M and N are the midpoints of $\overrightarrow{OA}$ and $\overrightarrow{OB}$ respectively.

EXAM PRACTICE QUESTION 4.161 (ANSWERS: (i) $\overrightarrow{MN}=\frac{a}{2}-\frac{b}{2}$, (ii) $\overrightarrow{NP}=\frac{b}{2}$ (iii) $\overrightarrow{PM}=-\frac{a}{2}$)

In the triangle OAB of the following figure, $\overrightarrow{OA}=\mathbf{a}$, $\overrightarrow{OB}=\mathbf{b}$ and M, N, P are the midpoints of $\overrightarrow{OB}$, $\overrightarrow{OA}$ and $\overrightarrow{AB}$ respectively. Find the vectors: (i) $\overrightarrow{MN}$, (ii) $\overrightarrow{NP}$, (iii) $\overrightarrow{PM}$ in terms of **a** and **b**

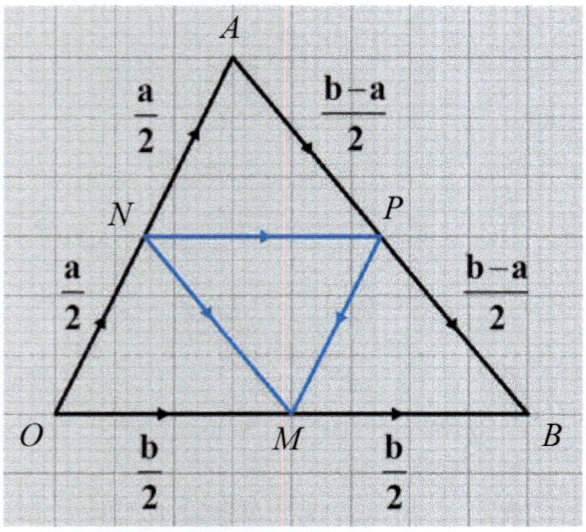

EXAM PRACTICE QUESTION 4.162 (ANSWERS: (i) $\overrightarrow{MN}=\frac{1}{2}(c-b)$, $\overrightarrow{BC}=c-b$ (ii) $\frac{\text{Area } \Delta ABC}{\text{Area } \Delta AMN}=4:1$)

In the adjacent figure, M and N are the midpoints of AB and AC respectively. If $\overrightarrow{AB}=\mathbf{b}$ and $\overrightarrow{AC}=\mathbf{c}$, express the vectors.

(i) $\overrightarrow{MN}$ and $\overrightarrow{BC}$ in terms of **b** and **c**.

(ii) Find the ratio of the area of the ΔABC to that of ΔAMN.

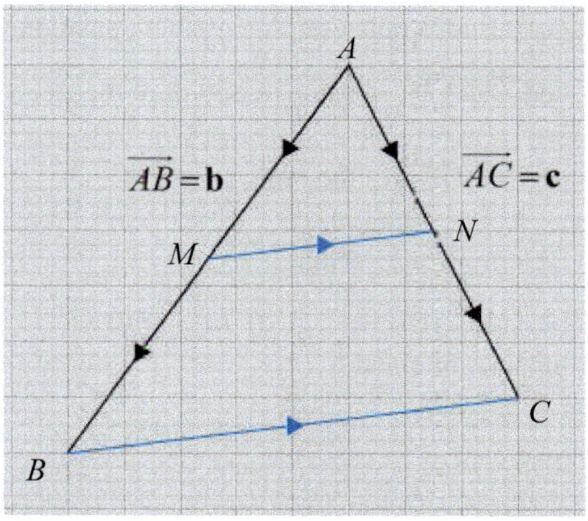

EXAM PRACTICE QUESTION 4.163 (ANSWERS: (i) $\overrightarrow{MC}=b-\frac{a}{2}$, $\overrightarrow{NA}=-\frac{1}{2}(a+b)$, $\overrightarrow{PB}=a-\frac{b}{2}$)

In a triangle ABC, M, N, P are the midpoints of AB, BC and AC respectively. MC, NA and PB are the medians.

If $\overrightarrow{AB}=\mathbf{a}$ and $\overrightarrow{AC}=\mathbf{b}$, find the vectors $\overrightarrow{MC}$, $\overrightarrow{NA}$ and $\overrightarrow{PB}$ in terms of **a** and **b**.

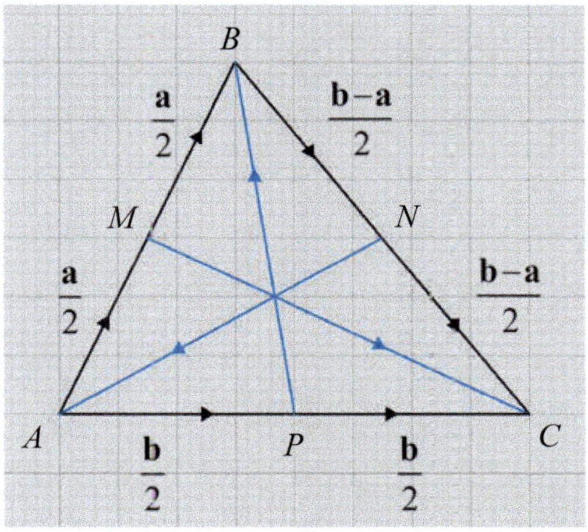

EXAM PRACTICE QUESTION 4.164

(ANSWERS: (i) $\overrightarrow{OM} = \frac{1}{2}(\mathbf{a}+\mathbf{b})$, (ii) $\overrightarrow{AP} = \frac{1}{3}\mathbf{b} - \frac{2}{3}\mathbf{a}$, (iii) $\overrightarrow{BP} = \frac{1}{3}\mathbf{a} - \frac{2}{3}\mathbf{b}$, (iv) $\overrightarrow{OP} = \frac{1}{3}(\mathbf{a}+\mathbf{b})$)

OAB is an equilateral triangle with position vectors $\overrightarrow{OA} = \mathbf{a}$ and $\overrightarrow{OB} = \mathbf{b}$. M is the midpoint of the free vector $\overrightarrow{AB}$ and P divides $\overrightarrow{OM}$ in the ratio $\frac{|\overrightarrow{OM}|}{|\overrightarrow{OP}|} = \frac{3}{2}$.

Determine the vectors,

(i) $\overrightarrow{OM}$, (ii) $\overrightarrow{AP}$, (iii) $\overrightarrow{BP}$, (iv) $\overrightarrow{OP}$
in terms of **a** and **b**

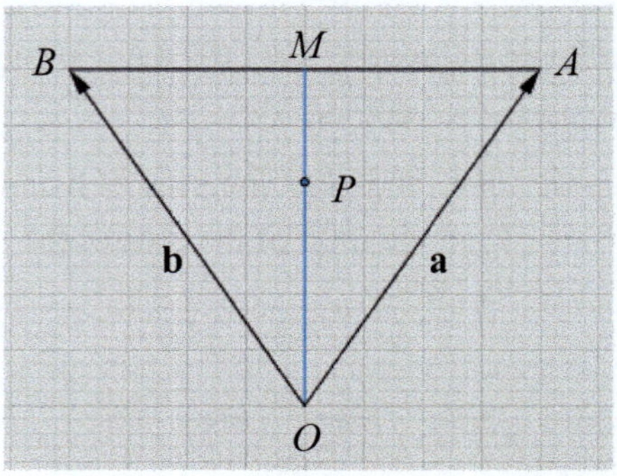

EXAM PRACTICE QUESTION 4.165

(ANSWERS: (a-i) $\overrightarrow{DC} = 3\mathbf{a}$, (a-ii) $\overrightarrow{BC} = 2\mathbf{a} - \mathbf{b}$, (a-iii) $\overrightarrow{BD} = -\mathbf{a} - \mathbf{b}$, (c) $\overrightarrow{PQ} = \frac{7}{3}\mathbf{a} - \frac{1}{3}\mathbf{b}$)

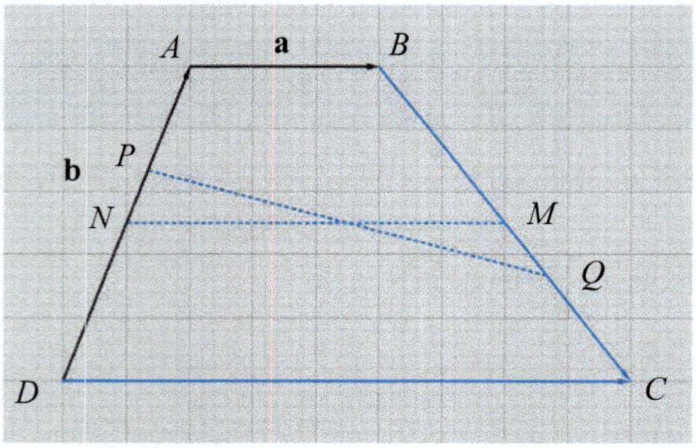

It is given that N is the midpoint of AD, DC is 3 times as long as AB and M is the midpoint of BC.

If $\overrightarrow{AB} = \mathbf{a}$ and $\overrightarrow{DA} = \mathbf{b}$,

(a) Find in terms of **a** and **b**:

(i) $\overrightarrow{DC}$ (ii) $\overrightarrow{BC}$ (iii) $\overrightarrow{BD}$

(b) show that NM is parallel to DC.

(c) If AP is one third distance of AD ($|\overrightarrow{AP}| = \frac{1}{3}\mathbf{b}$), and BQ is two thirds distance of BC ($|\overrightarrow{BQ}| = \frac{2}{3}|\overrightarrow{BC}|$), find $\overrightarrow{PQ}$.

EXAM PRACTICE QUESTION 4.166

(ANSWERS: (i) $\overrightarrow{MN} = \mathbf{a} + \mathbf{b}$, (ii) $\overrightarrow{NP} = \mathbf{b} + \mathbf{c}$, (iii) $\overrightarrow{MQ} = \mathbf{b} + \mathbf{c}$, (iv) $\overrightarrow{QP} = \mathbf{a} + \mathbf{b}$, (v) $\overrightarrow{MN} = \overrightarrow{QP} = \mathbf{a} + \mathbf{b}$ and $\overrightarrow{MQ} = \overrightarrow{NP} = \mathbf{b} + \mathbf{c}$ $\therefore$ *MNPQ* is a parallelogram)

ABCD is a quadrilateral and *M, N, P* & *Q* are the midpoints of *AB, BC, CD* and *DA* respectively. Given that $\overrightarrow{AM} = \mathbf{a}$, $\overrightarrow{BN} = \mathbf{b}$ and $\overrightarrow{CP} = \mathbf{c}$, find in terms of **a, b** and **c**, the vectors:

(i) $\overrightarrow{MN}$ (ii) $\overrightarrow{NP}$

(iii) $\overrightarrow{MQ}$ (iv) $\overrightarrow{QP}$

(v) Prove that *MNPQ* is a parallelogram.

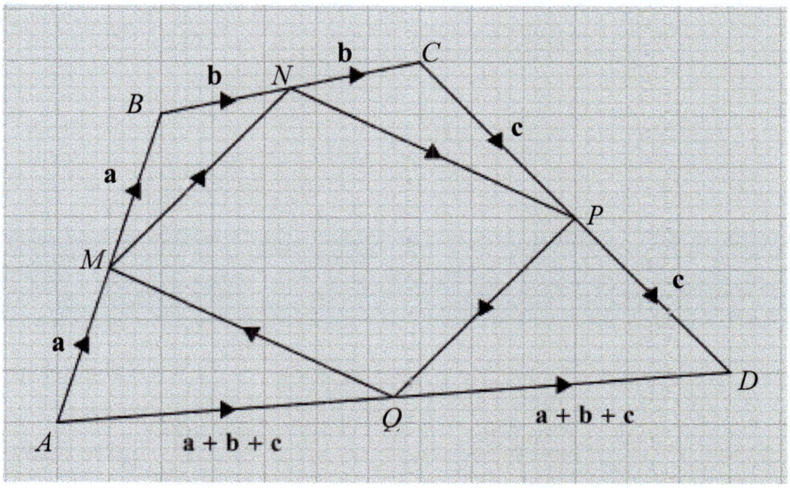

EXAM PRACTICE QUESTION 4.167

(ANSWER: $MN \parallel QP$ and $NP \parallel MQ$, hence *MNPQ* is a parallelogram)

ABCD is a quadrilateral. The midpoints of *AB, BC, CD* and *DA* are *M, N, P* and *Q* respectively.
Show that *MNPQ* is a parallelogram.

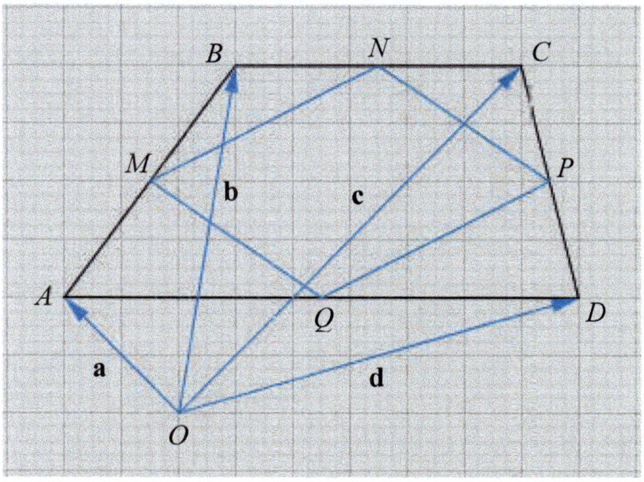

EXAM PRACTICE QUESTION 4.168 **(ANSWERS: (i) $\overrightarrow{AC} = \mathbf{a} + 2\mathbf{e}$, (ii) $\overrightarrow{BD} = \mathbf{e} - \mathbf{a}$, (iii) $\overrightarrow{CE} = -2\mathbf{a} - \mathbf{e}$, (iv) $\overrightarrow{EA} = \mathbf{a} - \mathbf{e}$,**

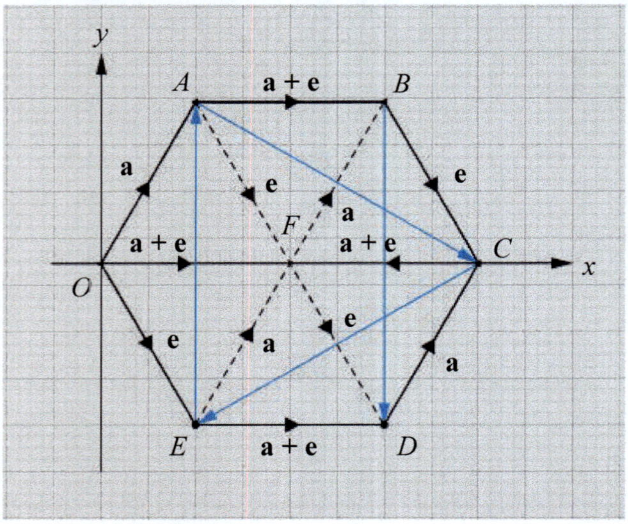

A regular hexagon *OABCDE*, with position vectors $\overrightarrow{OA} = \mathbf{a}$, $\overrightarrow{OE} = \mathbf{e}$ is shown in the figure below. Find the vectors:

(i) $\overrightarrow{AC}$, (ii) $\overrightarrow{BD}$, (iii) $\overrightarrow{CE}$

And (iv) $\overrightarrow{EA}$ in terms of **a** and **e**.

EXAM PRACTICE QUESTION 4.169

(ANSWERS: (i) $\overrightarrow{AB} = \mathbf{b} - \mathbf{a}$, (ii) $\overrightarrow{BC} = -\mathbf{a}$, (iii) $\overrightarrow{CD} = -\mathbf{b}$, (iv) $\overrightarrow{DE} = \mathbf{a} - \mathbf{b}$, (v) $\overrightarrow{EF} = \mathbf{a}$, (vi) $\overrightarrow{FA} = \mathbf{b}$

ABCDEF is a regular hexagon with diagonals intersecting at *O*, as shown in the following figure. If $\overrightarrow{OA} = \mathbf{a}$ and $\overrightarrow{OB} = \mathbf{b}$, find the vectors:

(i) $\overrightarrow{AB}$, (ii) $\overrightarrow{BC}$, (iii) $\overrightarrow{CD}$, (iv) $\overrightarrow{DE}$, (v) $\overrightarrow{EF}$, (vi) $\overrightarrow{FA}$ in terms of **a** and **b**.

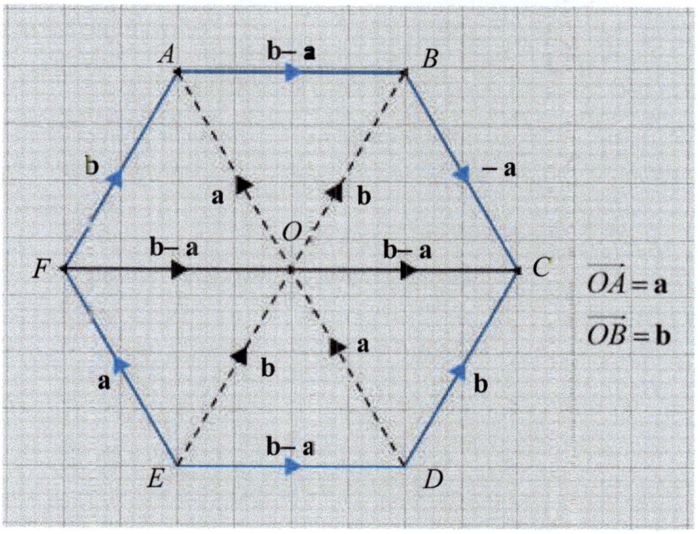

EXAM PRACTICE QUESTION 4.170

(ANSWER: 348 newton to 3 s.f., 55.7^0 to 3 s.f.)

Two forces of 100 N and 250 N are applied at a point *O*, at the direction of 45° and 60° from the horizontal axis *Ox*. Determine the resultant force and its direction from the horizontal axis. Use sine and cosine rules.

EXAM PRACTICE QUESTION 4.171 **(ANSWER: 140 newton to 3 s.f., 60.34^0 to 2 d.p.)**

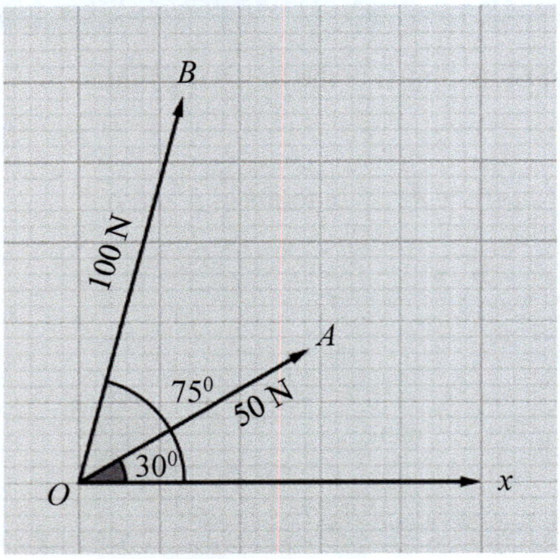

Two forces of 50 N and 100 N are applied at a point at directions 30° and 75° from the horizontal axis *Ox* respectively.

Determine the resultant force $\overrightarrow{OC}$ and its direction from the horizontal axis *Ox*.

Hint: Find first the angle $O\hat{A}C$ and then use the cosine rule to find $\overrightarrow{OC}$. Next, use the sine rule to find the direction of $\overrightarrow{OC}$. to the horizontal.

Topic 4.18 BEARINGS.

EXAM PRACTICE QUESTION 4.172 (ANSWERS: (i) 065^0 (ii) 245^0)

In the adjacent figure, determine the bearings (i) from *A* to *B* and (ii) from *B* to *A*.

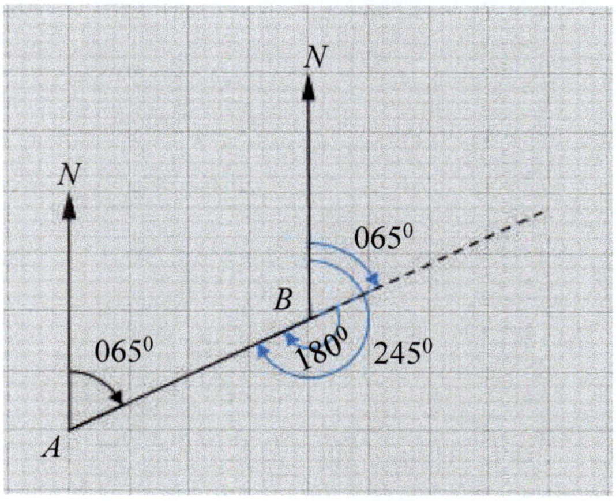

EXAM PRACTICE QUESTION 4.173 (ANSWER: 16.5 km to 3 s.f.)

At a given time t, the 3-digit bearings of two speed boats *A* and *B* from an observation point are 043° and 123° respectively, as shown in the adjacent figure. If *OA* = 10 km and *OB* = 15 km, find the distance *AB*.

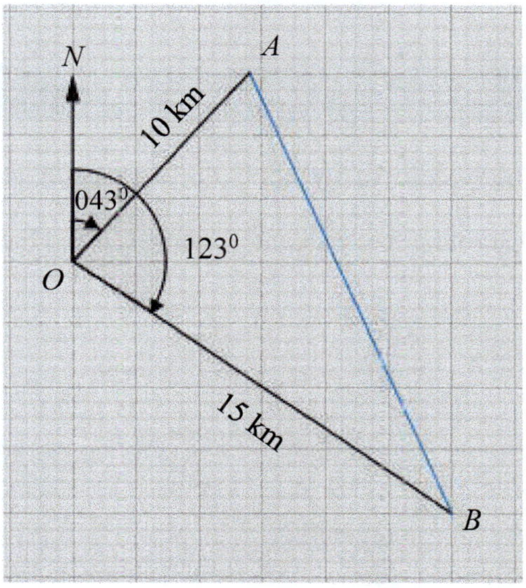

EXAM PRACTICE QUESTION 4.174 (ANSWER: 7.37 km/h, 106.3^0)

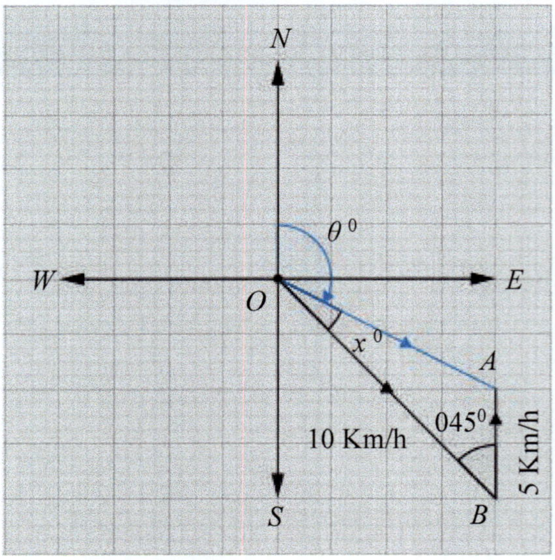

A yacht is sailing at 10 km/h in a S-E direction (line *OB*) and a tide of 5 km/h is running towards the N (line *BA*). Determine the velocity and direction of the yacht.
($\overrightarrow{OA}$ and θ)

EXAM PRACTICE QUESTION 4.175 (ANSWER:28.8 km/h to 3 s.f., 52.1^0 to 3 s.f.)

The velocity of a yacht is $\overrightarrow{OA} = 25$ km/h at 045° and tide of $\overrightarrow{AB} = 5$ km/h is running towards E.

Determine the resultant velocity $\overrightarrow{OB}$ and the bearing of the yacht.

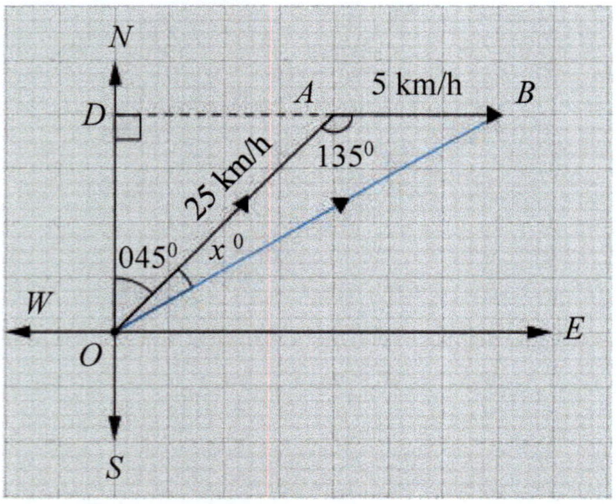

5 PROBABILITY

The difficulty level of the following exam-style questions and problems is progressively increasing and their full, step-by-step solutions with the required theoretical notes, diagrams and figures can be found in the book «GCSE (9-1) Mathematics Higher»[1], pp. 381-430. The numbering of the questions has remained the same, so, for example, the full solution of «Exam Practice Question 5.1» can be found in «Worked Example 5.1.» of the above book.

With this arrangement, it is simple for the students to read, understand and remember the theory and also to check and correct their work. For their convenience, we provide the numeric answers, so the students can check and correct their answers immediately. This is not possible when figures and diagrams are required. In such cases the students must refer to the corresponding Worked examples in «GCSE (9-1) Mathematics Higher»[1].

Topic 5.1 BASIC PROBABILITY–THEORETICAL PROBABILITY.

EXAM PRACTICE QUESTION 5.1 (ANSWER: $\frac{1}{13}$)

In a well shuffled pack of 52 cards where there are 4 suits, namely hearts, diamonds, clubs and spades, that is, 4 aces, 4 twos, 4 threes, 4 fours, 4 fives, 4 sixes, 4 sevens, 4 eights, 4 nines, 4 tens, 4 Jacks, 4 Queens and 4 Kings. The cards are either red or black. What is the probability of drawing an ace?

[1] Dr Gerasimos Masouros & Anthony Nicolaides, «GCSE (9-1) Mathematics Higher», PASS Publications Ltd, London 2018, ISBN 9781872684512

EXAM PRACTICE QUESTION 5.2 (**ANSWER: for a pack with no jokers (i) $\frac{1}{26}$, (ii) $\frac{1}{26}$, (iii) $\frac{1}{52}$, (iv) $\frac{1}{52}$, for two jokers in the pack (i) $\frac{1}{27}$, (ii) $\frac{1}{27}$, (iii) $\frac{1}{54}$, (iv) $\frac{1}{54}$)**

In the above pack of 52 cards, what is the probability for drawing:
(i) a black ace, (ii) a red ace, (iii) an ace of diamonds, (iv) an ace of clubs.
Recalculate the above if the pack of cards contains two jokers.

EXAM PRACTICE QUESTION 5.3 (**ANSWER: P(yellow) = 0.65)**

A bag contains red, blue and yellow balls. The probability of choosing a red ball is 0.25 and the probability of choosing a blue ball is 0.1. Workout the probability of choosing a yellow ball.

EXAM PRACTICE QUESTION 5.4 (**ANSWER: (i) girls' row: 8, 10, 18, boys' row: 7, 5, 12, totals' row 15, 15, 30, (ii) P(girl without glasses = $\frac{1}{3}$)**

The following two-way table gives the numbers of boys and girls in year 10 in a certain school, who wear glasses. (i) Complete the table:

	wear glasses	do not wear glasses	Total
Girls	8		
Boy		5	12
Total			30

(ii) A student is picked at random.
What is the probability that the student is a girl without glasses.

EXAM PRACTICE QUESTION 5.5 (ANSWER: (i) $P(A')=\frac{1}{3}$, (ii) $P(A')=\frac{1}{5}$, (iii) $P(A')=\frac{5}{6}$, (iv) $P(A')=0.25$, (v) $P(A')=0.08$)

Assume that the probability of a particular result A happening is:

(i) $\frac{2}{3}$ (ii) $\frac{4}{5}$ (iii) $\frac{1}{6}$ (iv) 0.75 (v) 0.92

Find the probabilities of the result not happening, or equivalently, find the probability of the complement of A.

EXAM PRACTICE QUESTION 5.6 (ANSWER: P(scoring a penalty goal = 0.985)

The probability of a football player missing a penalty goal is 0.015. What is the probability of scoring a goal.

EXAM PRACTICE QUESTION 5.7

(ANSWER: (a) P(*mango*) = 0.3, (b) P(*not mango*) = 0.7, there are 20 sweets with orange flavour)

A bag of 100 sweets is in three flavours: orange, mango, pineapple. The following table gives the probabilities of the flavours in this bag:

Flavour	Orange	Mango	Pineapple
P(flavour)	0.2		0.5

Work out the probabilities (a) P(*mango*), (b) P(*not mango*) and calculate the number of sweets of the orange flavour.

EXAM PRACTICE QUESTION 5.8

(ANSWER: 185 students are expected to choose the "Learning a musical instrument")

In a school of 1 000 students, each one had to choose one extra-curricular from the list below:
(a) Learning a foreign language.
(b) Learning a musical instrument.
(c) Playing a game.
(d) Participating in a field event.
(e) Learning computer programming.
The following tables gives the probability P, for each different choice.

Extra-curricular	Foreign Language	Musical Instrument	Game	Field Event	Computer Programming
P	0.015	x	0.1	0.45	0.25

How many students are expected to choose the learning of a musical instrument?

Topic 5.2 EXPERIMENTAL PROBABILITY.

EXAM PRACTICE QUESTION 5.9 **(ANSWER: P(R) = 0.55, P(B) = 0.45, with more trials, we get results which are closer to the theoretical probability)**

A roulette ball was spun forty times. The frequency (the number of outcomes) of red and black pockets were recorded as follows:

1	2	3	4	5	6	7	8	9	10	11	12	13	14	15	16	17	18	19	20
	R	R	R	R		R			R	R		R				R	R	R	
B					B		B	B			B		B	B	B				B

21	22	23	24	25	26	27	28	29	30	31	32	33	34	35	36	37	38	39	40
R		R				R	R		R	R	R	R	R	R	R				
	B		B	B	B			B								B	B	B	B

Workout the experimental probabilities P(R) and P(B).
Comment on the results.

EXAM PRACTICE QUESTION 5.10 **(ANSWER: P(H) = 0.4, If the coin was tossed 500 times, P(H) would have been closer to 0.5)**

A fair coin is tossed one hundred times and the heads and tails were recorded. 60 tails and 40 heads were recorded.
Workout the experimental probability for obtaining a Head. If the same coin was tossed 500 times, what would have been the probability of obtaining a head then?

Topic 5.3 SAMPLE SPACE DIAGRAM.

EXAM PRACTICE QUESTION 5.11 **(ANSWER: all possible outcomes of throwing two fair dice, are shown on the ample space diagram. P(7) = $\frac{1}{6}$)**

Find the sum of throwing two fair dice by listing all the possible outcomes. Workout the probability of obtaining a sum of 7.

EXAM PRACTICE QUESTION 5.12 **(ANSWER: P(*total 9*) = $\frac{1}{9}$, P(*both dice are the same*) = $\frac{1}{6}$), P(*difference of 3*) = $\frac{1}{6}$)**

Two fair dice are thrown. Using a sample space diagram, workout the probability of scoring:

(a) a total of 9 (b) both the same (c) the difference of 3.

EXAM PRACTICE QUESTION 5.13 (ANSWER: (ii-a) P(*at least 2T*) = $\frac{1}{4}$, (ii-b) P(*at least 2T*) = $\frac{4}{8}$, (ii-c) P(*at least 2T*) = $\frac{11}{16}$, (iii-a) P(*exactly 2T*) = $\frac{1}{4}$, (iii-b) P(*exactly 2T*) = $\frac{3}{8}$, (iii-c) P(*exactly 2T*) = $\frac{3}{8}$, (iv-a) P(*2T in the row*) = $\frac{1}{4}$, (iv-b) P(*2T in the row*) = $\frac{3}{8}$, (iv-c) P(*2T in the row*) = $\frac{1}{2}$)

Two, three and four fair coins are tossed in turn and the possible results of each coin are 'heads' (H) or 'tails' (T).

(i) List all the possible outcomes for each different experiment.

(ii) In each different experiment, find the probability, that at least 2 tails are tossed.

(iii) In each different experiment, find the probability, that exactly 2 tails are tossed.

(iv) In each different experiment, find the probability, that 2 tails in the row are tossed.

Topic 5.4 SET THEORY.

EXAM PRACTICE QUESTION 5.14 (ANSWER: (a) No, (b) No, (c) No, (d) Yes, (e) Yes)

Check whether the following sets are well defined or not:

(a) $A = \{x : \text{the old people of a village}\}$

(b) $B = \{\text{the beautiful animals of the zoo}\}$

(c) $C = \{x : \text{heavy packages}\}$

(d) $D = \{x : \text{the capital city of France}\}$

(e) $E = \{x : \text{positive integers from 1 to 5, 1 and 5 are included}\}$

EXAM PRACTICE QUESTION 5.15

Redraw the given Venn diagram and shade:

(i) $C = A \cap B$ (ii) $A \cup B$

(iii) $A' \cap B$ (iv) $A \cap B'$

(v) $A' \cap B'$ (vi) $(A \cup B)'$

(vii) $A \cup C$ (viii) $B \cup C$

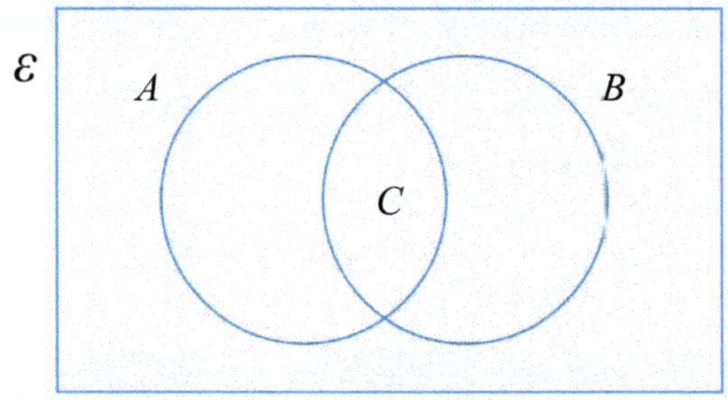

(i) Shade the area $C = A \cap B$

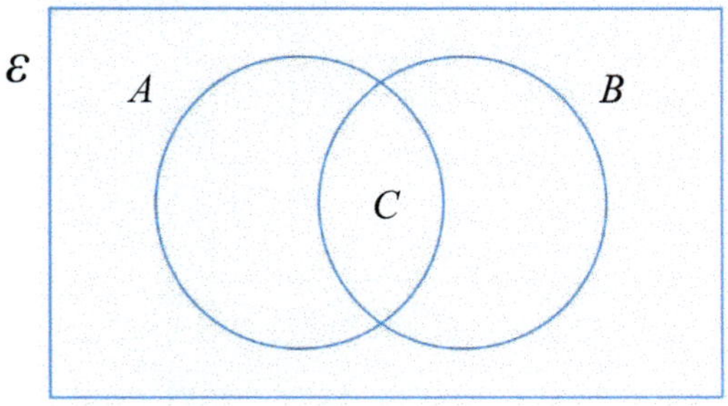

(ii) Shade the area $A \cup B$

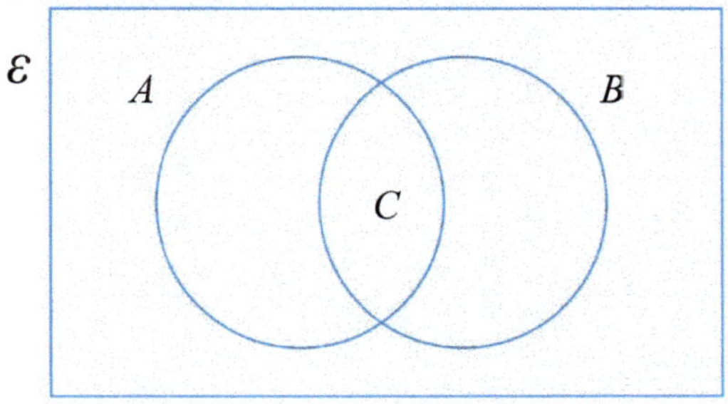

(iii) Shade the area $A' \cap B$

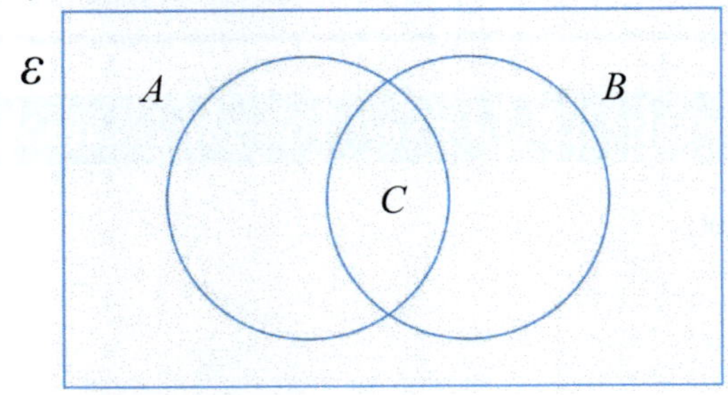

(iv) Shade the area $A \cap B'$

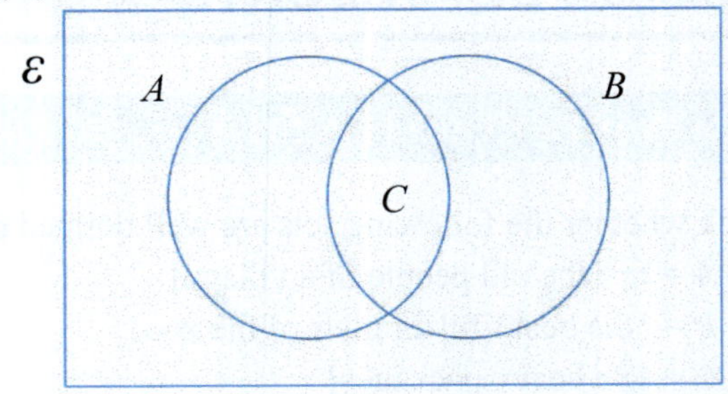

(v) Shade the area $A' \cap B'$

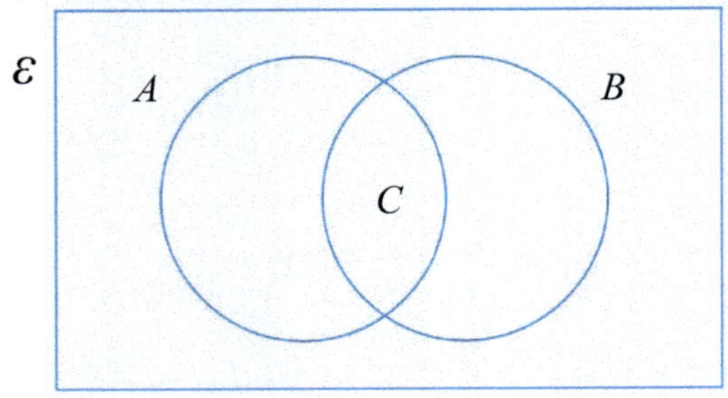

(vi) Shade the area $(A \cup B)'$

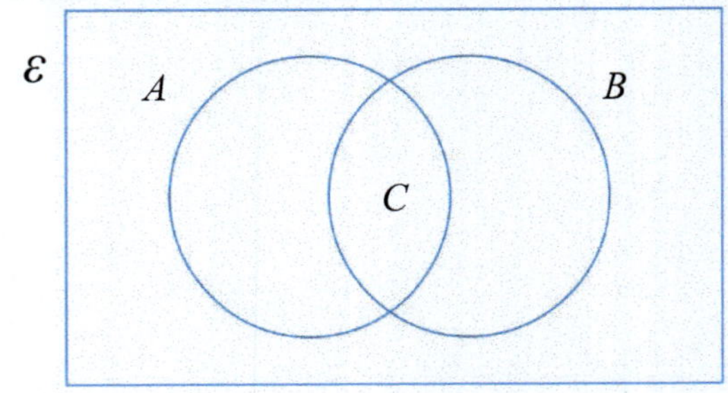

(vii) Shade the area $A \cup C = A$

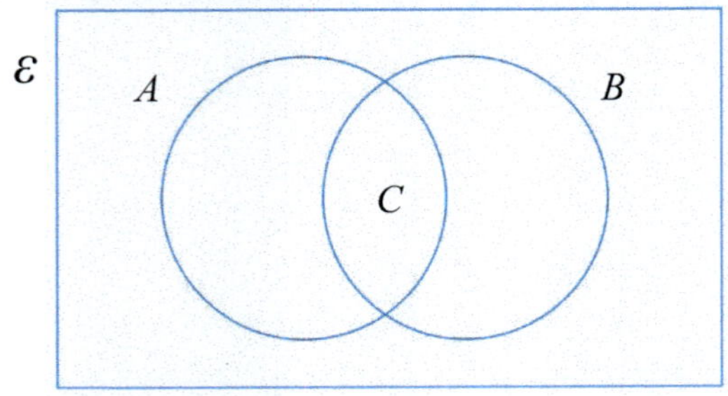

(viii) Shade the area $B \cup C = B$

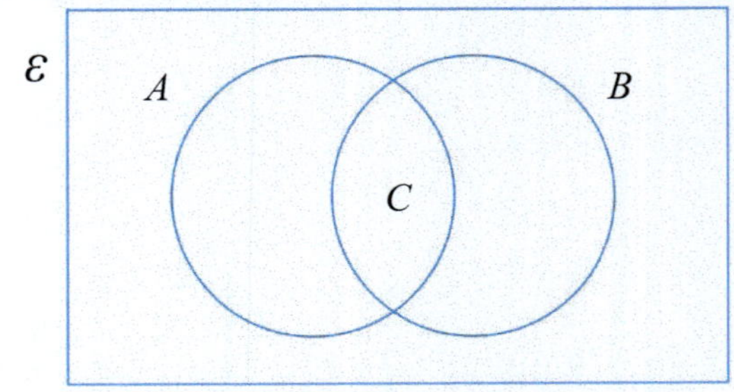

EXAM PRACTICE QUESTION 5.16 (ANSWER: true, false, false)

For the given set $A = \{a, b, c, d\}$, which of the following are true and why?
$a \in A$, $J \in A$, $c \notin A$.

EXAM PRACTICE QUESTION 5.17

Represent graphically the following sets:

(i) $A = \{x :$ days of week that begin with $S\}$
(ii) $B = \{x :$ a multiple of 5 between 0 and 41$\}$
(iii) $C = \{x, 5, y\}$

EXAM PRACTICE QUESTION 5.18 **(ANSWER: remember that the empty set and the given set itself are two subsets of any given set)**

(a) Write a subset of the Universal Set of the prime numbers which are smaller than 20.

(b) Write down the set P_{100} of all the prime numbers from 1 to 100.

EXAM PRACTICE QUESTION 5.19 **(ANSWERS: (a) $(A \cup B) = \mathcal{E}$, (ii) $C = \varnothing$ and $D = \mathcal{E}$)**

(a) If $(A \cup B)' = \varnothing$, what does this tell you about A and B?

(b) If $C \cup D' = \varnothing$, what does this tell you about C and D?

EXAM PRACTICE QUESTION 5.20

Represent the following by Venn diagrams:

(a) The complement A' of a set A

(b) $A \subset B \subset C \subset \mathcal{E}$

(c) $A \cup B = \{a, b, c\} \cup \{d, e, f\}$

(d) $P \cup Q = \{1, 2, 3, 4\} \cup \{2, 3, 4, 5, 6\}$

(e) $A \cap B = \{1, 3, 7, 25\} \cap \{2, 3, 5, 11\}$

(f) $P \cap Q = \{1, 2, 3, 4\} \cap \{2, 3, 4, 5, 6\}$.

EXAM PRACTICE QUESTION 5.21 **(ANSWERS: (i) $3 \in \{2,3\}$, (ii) there are 25 elements in the set A, (iii) there are 15 elements in the set A, (iv) Yes, (v) Yes)**

Explain the following statements:

(i) $3 \in \{x : x \text{ is a prime factor of } 24\}$

(ii) If $A = \{1^2, 2^2, 3^2, \ldots, 25^2\}$ then $n(A) = 25$

(iii) If $A = \{2, 4, 6, 8, \ldots, 30\}$ then $n(A) = 15$

(iv) The subsets of $A = \{a, b, c, d\}$ are : $\emptyset$, $\{a\}$, $\{b\}$, $\{c\}$, $\{d\}$, $\{a, b\}$, $\{a, c\}$, $\{a, d\}$, $\{b, c\}$, $\{b, d\}$, $\{c, d\}$, $\{a, b, c\}$, $\{a, c, d\}$, $\{a, b, d\}$, $\{b, c, d\}$, $\{a, b, c, d\}$

(v) A set which has 7 elements, has 126 proper subsets.

EXAM PRACTICE QUESTION 5.22 **(ANSWERS: (b) $A \cup B = \{2, 3, 4, 5, 6, 7, 8, 10, 11, 12\}$, (c) $n(A \cap B) = 1$, (d) $n(A \cup B) = 10$)**

Assume that $\mathcal{E}$ = {all positive integers less than 13}, $A = \{2, 4, 6, 8, 10, 12\}$ and $B = \{2, 3, 5, 7, 11\}$.

(a) Illustrate this information on a Venn diagram.

(b) List $A \cup B$

(c) What is $n(A \cap B)$?

(d) What is $n(A \cup B)$?

EXAM PRACTICE QUESTION 5.23 **(ANSWERS: (a) $A' = \{0, 2, 3, 4, 6, 7, 9\}$, (b) $B' = \{0, 2, 4, 6, 10\}$, (c) $A \cup B = \{1, 3, 5, 7, 8, 9, 10\}$, (d) $A \cap B = \{1, 5, 8\}$)**

If $A = \{1, 5, 8, 10\}$, $B = \{1, 3, 5, 7, 8, 9\}$ and $\mathcal{E} = \{0, 1, 2, 3, 4, 5, 6, 7, 8, 9, 10\}$, draw a Venn diagram to represent these sets. Hence write down the elements of the following sets:
(a) A' (b) B' (c) $A \cup B$ (d) $A \cap B$.

EXAM PRACTICE QUESTION 5.24 **(ANSWER: $x = 8$)**

In a year 11 class (top set), there are 50 students. Some decided to take Mathematics only for A levels, some decided to take Statistics only, a few decided to take both and ten decided to take neither.
If 30 took Mathematics and 18 took Statistics, how many students took both subjects.

EXAM PRACTICE QUESTION 5.25 **(ANSWER: $x = 3$)**

In a class of 30 GCSE students, 20 take Mathematics, 8 take Statistics and 5 take neither subject. How many students take both subjects?

EXAM PRACTICE QUESTION 5.26

(ANSWER: $A \cup B = \{1, 2, 3, 5, 7, 9, 11, 13\}$, and $n(A \cup B) = 8$)

Given that, $\mathcal{E}$ = {all the positive integers less than 15}, A = {all positive prime numbers less than 15} and B = {all positive odd numbers less than 15}.

(a) Illustrate these sets on a Venn diagram.

(b) Find the set $A \cup B$ and $n(A \cup B)$.

EXAM PRACTICE QUESTION 5.27

(ANSWERS: (i) 4, (ii) 1, (iii) 1, (iv) 1, (v) 1, (vi) 9, (vii-i) {a, b, c, d}, (vii-ii) {b}, (vii-iii) {*f*}, (vii-iv) {c}, (vii-v) {*k*}, (vii-vi) {a, b, c, d, e, *f*, g, h, *k*})

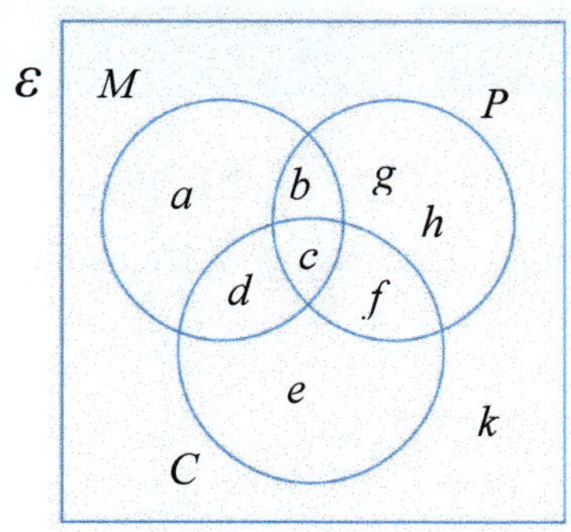

On the adjacent Venn diagram $\mathcal{E}$ = {names of people taking Examination}, M = {Mathematics}, P = {Physics Exams}, C = {Chemistry Exams}.

(i) How many people took Mathematics?

(ii) How many people took Mathematics and Physics only?

(iii) How many people took Physics and Chemistry only?

(iv) How many people took all three subjects?

(v) How many did not take any of these?

(vi) How many students were there?

(vii) Write the set of the elements for all the above.

EXAM PRACTICE QUESTION 5.28 **(ANSWERS: (i) {1, 2, 3, 5, 7, 9, 11, 13}, (ii) {3, 5, 7, 11, 13}, (iii) 8, (iv) 5, (v) ∅, (vi) {1, 2, 9})**

If $\mathcal{E} = (A \cup B)$, where $A = \{1, 3, 5, 7, 9, 11, 13\}$ is the set of odd numbers less than 14 and $B = \{2, 3, 5, 7, 11, 13\}$ is the set of prime numbers less than 14, draw the corresponding Venn diagram and find:

(i) $A \cup B$ (iii) $n(A \cup B)$ (v) $A' \cap B'$

(ii) $A \cap B$ (iv) $n(A \cap B)$ (vi) $(A \cup B)'$

EXAM PRACTICE QUESTION 5.29 **(ANSWERS: (a) {1, 2, 3, 4, 5, 6, 7, 8, 11, 12, 13, 17, 19, 23, 24, 29, 31}, (b) {2, 3}, (c) 11, (d) {9, 10, 14, 15, 16, 18, 20, 21, 22, 25, 26, 27, 28, 30, 32, 33, 34, 35, 36, 37, 38, 39, 40}, (e) {1, 4, 6, 8, 12, 24}, (f) 5, 7, 11, 13, 17, 19, 23, 29, 31})**

If $\mathcal{E}$ = {all the positive integers from 1 to 40}, $A = \{2, 3, 5, 7, 11, 13, 17, 19, 23, 29, 31\}$ and B = {factors of 24}, draw a Venn diagram to show the above information. Hence use your Venn diagram to find:

(a) $A \cup B$ (b) $A \cap B$ (c) $n(A)$

(d) $A' \cap B'$ (e) $A' \cap B$ (f) $A \cap B'$.

EXAM PRACTICE QUESTION 5.30 (ANSWERS: (a) $\varnothing$, (b) $\mathcal{E}$)

If $A=\{1, 3, 5, 7, 9\}$, $B=\{0, 2, 4, 6, 8, 10\}$ and $\mathcal{E}=\{0, 1, 2, 3, 4, 5, 6, 7, 8, 9, 10\}$, draw a Venn diagram to represent the sets. Hence find (a) $A \cap B$, (b) $A \cup B$.

EXAM PRACTICE QUESTION 5.31 (ANSWERS: (a) {2, 3, 5, 7, 11, 13, 17, 19}, (b) {3, 5, 13}, (c) {2, 23, 29, 31, 37, 41, 43, 47}, (d) {7,11, 17, 19, 23, 29, 31, 37, 41, 43, 47}, (e) {2}, (f) {7, 11, 17, 19},)

Let $\mathcal{E}$ = {All the prime numbers up to 50} = {2, 3, 5, 7, 11, 13, 17, 19, 23, 29, 31, 37, 41, 43, 47}.

A = {the elements of $\mathcal{E}$ which are odd numbers from 3 and including 21}.

B = {the elements of $\mathcal{E}$ which are numbers of Fibonacci sequence from 2 and including 13}.

Draw a Venn diagram to show the above information. Hence use your Venn diagram to find:

(a) $A \cup B$ (b) $A \cap B$ (c) A' (d) B' (e) $A' \cap B$ (f) $A \cup B'$.

Topic 5.5 PROBABILITY: SET NOTATION FOR ELEMENTARY EVENTS.

EXAM PRACTICE QUESTION 5.32 (ANSWER: Dependent and Not mutually exclusive)

Assume that $P(A_1) = 0.35$, $P(A_2) = 0.4$ and $P(A_1 \text{ OR } A_2) = 0.615$. What is the relationship of the events A_1 and A_2 (mutually exclusive, independent or none of the two).

EXAM PRACTICE QUESTION 5.33 (ANSWERS: (a-i) $\frac{1}{4}$, (a-ii) $\frac{1}{4}$, (a-iii) $\frac{1}{4}$, (a-iv) $\frac{1}{4}$, (a-v) $\frac{1}{2}$, (a-vi) $\frac{1}{2}$, (a-vii) $\frac{1}{2}$, (a-viii) $\frac{1}{2}$, (a-ix) $\frac{3}{4}$, (b-i) $\frac{4}{13}$, (b-ii) $\frac{4}{13}$, (b-iii) $\frac{4}{13}$, (b-iv) $\frac{4}{13}$)

In a pack of 52 playing cards, where C represents Clubs, D represents Diamonds, S represents Spades and H represents Hearts, we have $13C$, $13D$, $13S$ and $13H$. If we draw one and from this pack, find

(a) (i) $P(C)$, (ii) $P(D)$, (iii) $P(S)$, (iv) $P(H)$,
(v) $P(H \text{ or } C)$, (vi) $P(H \text{ or } D)$, (vii) $P(C \text{ or D})$, (viii) $P(S \text{ or } D)$ and
(ix) $P(C, D \text{ or } S)$

(b) Using the addition law, determine (i) P(D or a ten), (ii) P(C or Queen), (iii) P(S or King),
(iv) P(H or an eight)

EXAM PRACTICE QUESTION 5.34 (ANSWER: Prove that both conditions of the mathematical test are satisfied)

Tossing a coin we have two outcomes: Heads (H) OR Tails (T). Prove that the events H and T are mutually exclusive events.

EXAM PRACTICE QUESTION 5.35 (ANSWER: 0.25)

In a football match three events can happen. The team can win (W), the team can lose (L), or the team can draw (D). Assume that in a certain match, P(W) = Probability of winning = 0.40 and P(L) = Probability of losing = 0.35. Find the probability of drawing P(D).

EXAM PRACTICE QUESTION 5.36 (ANSWER: $\frac{2}{13}$)

A card is randomly selected from a pack of 52 playing cards. Find the probability of drawing an ace or a jack.

EXAM PRACTICE QUESTION 5.37 (ANSWER: $\frac{4}{13}$)

A card is randomly drawn from a pack of 52 playing cards. Find the probability of drawing an Ace (A) or a Heart (H).

EXAM PRACTICE QUESTION 5.38 (ANSWERS: (a) $\frac{1}{36}$, (b) $\frac{2}{36}$, (c) $\frac{6}{36}$)

A dice is thrown twice,

(a) Find the probability of throwing a "six" followed by a "five", that is a "six" on the first throw and a "five" on the second throw.

(b) Find the probability of throwing a "six" and a "five", that is a "six" on the first throw and a "five" on the second throw or a "five" on the first throw and a six on the second throw.

(c) Find the probability of throwing "doublets", that is (1,1), (2,2), (3,3) and so on.

EXAM PRACTICE QUESTION 5.39 (ANSWER: $\frac{1}{221}$)

Remove two cards from a pack of 52 playing cards, without replacement. Find the probability of drawing two kings (K).

EXAM PRACTICE QUESTION 5.40 (ANSWERS: (i) $\frac{1}{221}$, (ii) $\frac{1}{169}$)

From a pack of 52 playing cards we draw two successive cards (i) without replacing the first card (ii) with replacing the first card. Find the probability in each case of obtaining two Queens.

EXAM PRACTICE QUESTION 5.41 (ANSWERS: (a) $\frac{25}{102}$, (b) $\frac{13}{51}$)

From a pack of well shuffled 52 playing cards we draw one card at random and then another card without replacing the first card. Calculate the probability that:

(a) Both cards are black.

(b) The first card is red and the second card is black.

EXAM PRACTICE QUESTION 5.42 (ANSWER: they are not mutually exclusive)

When rolling a fair dice, the event of obtaining a prime number is $A_1 = \{2, 3, 5\}$. The event of obtaining an even number is $A_2 = \{2, 4, 6\}$. Are the events A_1 and A_2 mutually exclusive?

EXAM PRACTICE QUESTION 5.43 (ANSWER: $\frac{1}{10}$)

In a year 11 class there are 30 students, 15 students work on homework A, 8 work on homework B and 10 work on neither of these homeworks. A student is chosen at random. Workout the probability that the student works on both homeworks.

EXAM PRACTICE QUESTION 5.44 (ANSWERS: (a) $\frac{1}{7}$, (b) $\frac{2}{7}$, (c) $\frac{2}{7}$)

A box contains 4 red marbles and 3 white marbles. We draw a marble at random and we note its colour and do not replace it, then we draw another marble. What is the probability

(a) The two marbles are white,

(b) The two marbles are red,

(c) The first is red and the second is white.

EXAM PRACTICE QUESTION 5.45 (ANSWERS: (a) 0.650, (b) 0.275, (c) 0.444, (d) 0.043)

The highest score on a dart board with one dart is 60. The probability that John will hit this region is 0.35. John has 3 darts. Assuming that John is not making any adjustments to his next throw, based on the outcome of his previous throw, workout the following probabilities:

(a) The probability that John does not hit the highest score with one dart

(b) The probability that John does not hit the highest score with 3 darts

(c) The probability that John uses 3 darts but hits the highest score with only one dart

(d) What is the probability that John hits the high score with all 3 darts?

Give your answers to 3 d.p.

EXAM PRACTICE QUESTION 5.46 (ANSWER: $\frac{1}{216}$)

Three fair dice, A, B, and C are thrown on to a horizontal table. What is the probability of obtaining three sixes.

Topic 5.6 TREE DIAGRAMS.

EXAM PRACTICE QUESTION 5.47

(ANSWERS: (a) $\frac{1}{36}$, (b) $\frac{5}{36}$)

What is the probability of rolling a dice for a second time, in order to obtain a 6, given that:

(a) 6 was obtained in the first roll.

(b) 6 was not obtained in the first roll.

Draw the tree diagram of the two rolls.

EXAM PRACTICE QUESTION 5.48 (ANSWER: $\frac{25}{72}$)

In a bag you have 5 red balls (R), 4 green balls (G) and 3 yellow balls (Y). All the balls have identical size. You select a ball, record the colour, put it back and repeat the experiment. What is the probability of the two balls being the same colour?

EXAM PRACTICE QUESTION 5.49 (ANSWER: $\frac{19}{66}$)

Repeat the previous example but do not replace the first pick.

EXAM PRACTICE QUESTION 5.50 (ANSWERS: (a) $\frac{7}{15}$, (b) $\frac{8}{15}$)

10 identical balls are in a bag. 6 balls are black and 4 are red. We pick one ball and do not replace it and then we pick the second ball. Show these picks by using a Tree Diagram and find the probability:

(a) To draw two balls of the same colour.

(b) To draw two balls of different colours

EXAM PRACTICE QUESTION 5.51 (ANSWERS: (a) $\frac{3}{10}$, (b) $\frac{7}{40}$)

A bag contains 10 identical marbles. Three marbles are red and the rest are yellow. A marble is removed from the bag and its colour is noted. This marble is not replaced in the bag. A second marble is picked and the process is repeated by a third marble.

Draw a tree diagram and show all the outcomes of the AND rule. Workout the probabilities

(a) of a matching colour

(b) of selecting exactly 2 reds.

Exam Practice Question 5.52 (Answers: (a) $\frac{1}{12}$, (b) $\frac{1}{36}$, (c) $\frac{1}{6}$, (d) $\frac{5}{18}$)

A bag contains 9 identical discs, 2 are red, 3 are yellow and 4 are blue. We draw a disc at random and we note its colour and do not replace it, then we draw another disc.
Workout the probabilities

(a) The first disc is red and the second one is yellow
(b) The two discs are red
(c) The two discs are blue
(d) A matching colour.

Illustrate all the outcomes using a tree diagram.

EXAM PRACTICE QUESTION 5.53 (ANSWERS: (c) $\frac{1}{12}$)

A coin is tossed and a dice is rolled. The coin can land on T = Tail or H = Head and the dice on 1, 2, 3, 4, 5 or 6.

(a) Draw a sample space diagram to show the probabilities of all the outcomes.

(b) Draw a tree diagram to illustrate all the outcomes with their probabilities.

(c) Work out the probability that a T is tossed and a 2 is rolled.

Exam Practice Question 5.54

(Answers: (a) $\frac{10}{n+10}$, (b) $\frac{n^2+100}{(n+10)^2}$, (c) 10, (d) $\frac{90+n(n-1)}{(n+10)(n+9)}$)

A bag contains n red and 10 black counters. A counter is chosen at random from the bag, its colour is noted and it is then replaced.

(a) Write down an expression in terms of n for the probability that the counter is black.

(b) A second counter is chosen at random. Work out the probability for a matching pair in two draws with replacement, in terms of n.

(c) Determine n, if the colours of the two counters are different with probability 0.5.

(d) Repeat (b) without replacement.

6 STATISTICS

The difficulty level of the following exam-style questions and problems is progressively increasing and their full, step-by-step solutions with the required theoretical notes, diagrams and figures can be found in the book «GCSE (9-1) Mathematics Higher»[1], pp. 431-496. The numbering of the questions has remained the same, so, for example, the full solution of «Exam Practice Question 6.1» can be found in «Worked Example 6.1.» of the above book.

With this arrangement, it is simple for the students to read, understand and remember the theory and also to check and correct their work. For their convenience, we provide the numeric answers, so the students can check and correct their answers immediately. This is not possible when figures, diagrams and answers to theoretical questions are required. In such cases the students must refer to the corresponding Worked examples in «GCSE (9-1) Mathematics Higher»[1].

Topic 6.1 INTRODUCTION.

EXAM PRACTICE QUESTION 6.1 **(ANSWER: Mathematics 6 papers, Physics 9 papers, Chemistry 5 papers)**

Probability Proportional to size sampling: In year 11, students offered Mathematics, Physics or Chemistry homework. 280 students chose Mathematics homework, 470 chose Physics homework and 250 chose Chemistry homework.
When the papers came back, if you wanted to check a sample of 20 papers, based on probability proportional to size sampling, how many of each subject would you choose?

[1] Dr Gerasimos Masouros & Anthony Nicolaides, «GCSE (9-1) Mathematics Higher», PASS Publications Ltd, London 2018, ISBN 9781872684512

EXAM PRACTICE QUESTION 6.2

Determine the population, the sample, the variables and the types of variables in the following instances:

(a) The preferences of the electors in sight of the forthcoming elections:

Population : ....

Sample :..............................

Variable :............................

Type of Variable : ....

(b) The height and the mass of year 10 and 11 students:

Population : ....

Sample :..............................

Variable :............................

Type of Variable : ....

(c) The blood group and the gender of year 10 and 11 students:

Population : ....

Sample :..............................

Variable :............................

Type of Variable: ....

(d) The consequences of smoking to people's health:

Population : ....

Sample :..............................

Variable :............................

Type of Variable : ....

(e) The salary of employees in a supermarket:

Population :.........................

Sample : ....

Variable : ....

Type of Variable :...............

(f) The number of customer complaints in a bank:

Population :.........................

Sample : ....

Variable: ....

Type of Variable: ....

EXAM PRACTICE QUESTION 6.3

Jack buys a new pair of trousers. Illustrate two: (a) qualitative (b) discrete variables, on the trousers.

(a) Qualitative data times: 1.

2.

(b) Discrete data types: 1.

2.

EXAM PRACTICE QUESTION 6.4

Which of the following variables are qualitative and which are quantitative?

(a) The number of shoe sizes for women :.....

(b) Weight : ....

(c) Profession : ....

(d) Birth place :..

EXAM PRACTICE QUESTION 6.5

Determine the types of variables in the following statements

(a) There are 40 orange trees in an orchard :..............

(b) There are several species of orange trees :............

(c) The diameter of the orange tree trunk :.................

(d) Number of oranges per tree :.................................

(e) Weight of the oranges per tree : ....

(f) Colour of the orange : ....

EXAM PRACTICE QUESTION 6.6

The sizes of the shoes of adult men are $8, 8\frac{1}{2}, 9, 9\frac{1}{2}, 10, 11$ and 12. Answer the questions:

Is the shoe size a discrete or a continuous variable (YES/NO, why)?

Is the length of a shoe a discrete or a continuous variable (YES/NO, why)?

EXAM PRACTICE QUESTION 6.7

For each one of the following cases, write at least two variables that would be of interest. Distinguish between qualitative and quantitative variables and mention any further classification (discrete/continuous).

(a) To examine a sample of employees of a company.

(b) To examine a sample of products from a production.

(c) To examine basketball players of a team in a match.

Topic 6.2 TALLY MARKS AND CHARTS.

Exam Practice Question 6.8

(a) Using tally marks, write the following numbers:

Numbers	Tally marks	Numbers	Tally marks	Numbers	Tally marks
17		24		15	
21		11.			

(b) Write the numbers corresponding to the following tally marks:

Tally marks	Numbers	Tally marks	Numbers	Tally marks	Numbers
𝍸 𝍸 ||||		𝍸 ||||		𝍸 𝍸 𝍸 |	
|||		𝍸 𝍸 𝍸			

Exam Practice Question 6.9 (Answers: (a) plain, (b) 40)

Jack carried a survey of his friends' favourite flavour of crisps.
The following 4 flavours were given:

(i) Cheese and Onion.
(ii) Salt and Vinegar.
(iii) Plain.
(iv) Chicken flavour.

The results are shown in the table below.

Flavour of crisps	**Tally**	**Number of friends (Frequency f_i)**
Cheese and Onion	𝍸 𝍸 |	11
Salt and Vinegar	𝍸 |||	8
Chicken	𝍸 |	6
Plain	𝍸 𝍸 𝍸	15

(a) Write down the favourite flavour of most of Jacks's friends.

(b) Write down the total number of friends.

Topic 6.3 SCATTER DIAGRAMS AND CORRELATIONS. INTERPOLATION AND EXTRAPOLATION.

EXAM PRACTICE QUESTION 6.10 (ANSWERS: (a) negative, (c) £1 700, (d) £1 430)

Plot the following pairs of points representing the age of a car and its corresponding price.

Ages (years)	0.5	1.2	6.1	3.9	5.2	6.7	2	3.1
Price (£)	1 950	1 800	1 540	1 740	1 680	1 530	1 810	1 970

(a) Describe the relationship between these variables.

(b) Draw the line of best fit and determine its gradient.

(c) Estimate the value of the 4 years old car (interpolation).

(d) Estimate the value of the 8 years old car (extrapolation).

Topic 6.4 MEAN, MEDIAN, MODE, RANGE. GROUPED DATA AND DIFFERENT TYPES OF FREQUENCIES.

Ungrouped Data

EXAM PRACTICE QUESTION 6.11 (ANSWERS: (a) 3.875, (b) 4, (c) 5, (d) 10)

In the set of integer values: 5, 9, −1, 0, 8, 3, 2, 5, find

(a) The mean $\overline{x}$ =

(b) The median =

(c) The mode =

(d) The range =

EXAM PRACTICE QUESTION 6.12 (ANSWERS: (a) 202 cm, (b)225 cm)

The mean height of five basketball players of a team is 201 cm.

(a) In order to raise the height of the team the trainer took one more player with height 207 cm. What is now the mean height of the team.

(b) If the trainer wanted to increase the height of the team to 205 cm what would be the height of the additional player.

EXAM PRACTICE QUESTION 6.13 (ANSWER: (a) 10, 12, 14, 16, 18, 20, (b) 15)

Six consecutive even numbers have a mean value of 15.
(a) Work out these even numbers and (b) find the median.

EXAM PRACTICE QUESTION 6.14 (ANSWERS: (a) 13, 15, 17, 19, 21, 23, 25, 27, (b) 20)

8 consecutive odd numbers have a mean value of 20.
(a) Work out these odd numbers and (b) find their median.

EXAM PRACTICE QUESTION 6.15 (ANSWER: 2, 6)

The mean value and the median of five numbers is 6. The three of the numbers are 5, 8, 9. Find the other 2 numbers.

EXAM PRACTICE QUESTION 6.16 (ANSWER: 4, 12)

The mean value and the median of three numbers is 12. One of the numbers is 20. Find the missing numbers.

EXAM PRACTICE QUESTION 6.17 (ANSWER: 14.8 to 3 s.f.)

The mean age of 14 boys and 16 girls of the mathematics class room is 15 years. If the mean age of the boys is 15.2 years, find the mean age of the girls to 3 significant figures.

EXAM PRACTICE QUESTION 6.18

(ANSWERS: (i) 60, (ii) 58, 60, 65, 75, (iii) 80, (iv) 59.7 to 3 s.f., (v) 52.5, (vi) 70, (vii) 17.5, (viii) 15)

A test in Mathematics was given in a class of 30 students and the marks out of 100 were as follows:

29	50	55	59	15	80	75	64	75	60	49	58	57	62	65
68	65	60	69	45	58	55	58	70	40	75	39	95	72	70

Workout:

(i) The median =

(ii) The modes =

(iii) The range =

(iv) The mean =

(v) The lower quartile =

(vi) The upper quartile =

(vii) The inter quartile range =

(viii) Are there any outliers?

EXAM PRACTICE QUESTION 6.19

(ANSWERS: (a) 7, (b) 6.3, (c) 9, 7, (d) 7, (e) 4, (f) 8.5, (g) 4.5, (h) NO)

20 students took a Mathematics test and obtained the following marks out of 10:

4	9	7	6	7	9	8	7	7	6	3	2	3	9	7	6	4	4	9	9

Workout the following:

(a) The median =

(b) The mean =

(c) The mode =

(d) The range =

(e) The lower quartile =

(f) The upper quartile =

(g) The inter quartile range =

(h)Are there any outliers?

EXAM PRACTICE QUESTION 6.20 (ANSWERS: (a) 7.42, (b) 6, (c) 7, (d) 6, (e) 9, (f) 3)

50 students took a chemistry test. The table below gives information about their marks out of 10 in the test.

Marks	Frequency (v_i)
4	4
5	6
6	3
7	15
8	4
9	10
10	8
	$\sum v_i = 50$

Workout the following:
(a) The mean mark
(b) The range
(c) The median
(d) The lower quartile
(e) The upper quartile
(f) The interquartile range

(a) The mean mark =

(b) The range =

(c) The median =

(d) The lower quartile =

(e) The upper quartile =

(f) The The inter quartile range =

EXAM PRACTICE QUESTION 6.21 (ANSWERS: (a) 50, 55, (b) 27.5, 67.5, (c) 40)

The marks out of 100 of 10 students in a test are: 35, 55, 50, 65, 75, 15, 60, 55, 20, 70
(a) Find the mean value and the median.
(b) Find the lower and upper quartiles.
(c) The interquartile range.

Grouped Data

EXAM PRACTICE QUESTION 6.22

The following table presents the heights in cm of 28 women, which are measured during an experiment.

170	175	167	178	178	180	179
180	175	177	165	173	189	180
165	173	180	178	187	176	170
170	160	182	175	170	167	173

Arrange the measurements in 5 classes of equal width.

Class width [ ]	Central values x_i	Tally	Absolute Frequencies v_i	Relative Frequencies $f_i\%$	Cumulative Absolute Frequencies N_i	Cumulative Relative Frequencies $F_i\%$
		Total				

EXAM PRACTICE QUESTION 6.23

(ANSWERS: (b) 13 or 26%, (c) 37 or 74%, (d) 7 or 14%, (f) 7, (g) 6, (h) 21, (i) 16, (j) 3, (k) 4)

A factory is producing a special part for medical equipment. 50 employees, men (M) and women (W), were observed and the numbers of parts they produced were recorded as follows:

3	M	4	M	5	W	8	M	9	W	8	M	6	M	8	M	7	W	10	W
8	W	7	W	6	M	5	M	9	M	3	W	8	M	5	W	6	M	6	M
6	M	3	M	5	M	6	M	4	W	2	W	9	M	8	W	7	M	7	W
1	W	6	W	3	W	1	M	5	W	8	W	1	M	2	M	3	M	4	W
5	M	6	M	7	W	9	M	10	W	9	W	8	M	7	M	6	M	5	W

The minimum required daily production for each employee is 5 parts, while anybody who produces 9 or more gets a bonus

(a) Draw a table to show the distribution of frequencies (absolute and cumulative) for all the employees, regardless their gender.

From the above table show:

(b) How many employees failed to produce 5 or more items (absolute number and percentage).

(c) How many employees produced 5, or more items (absolute number and percentage).

(d) How many employees won a bonus (absolute number and percentage).

(e) Draw another table to show the distribution of frequencies (absolute only) for Men and Women. How many men and how many women were chosen in the sample? Was the sample unbiased?

From the table in (e) show:

(f) How many Men failed to produce 5 items (absolute number only).

(g) How many Women failed to produce 5 items (absolute number only).

(h) How many Men produced 5, or more items (absolute number only).

(i) How many Women produced 5, or more items (absolute number only).

(j) How many Men won a bonus (absolute number only).

(k) How many Women won a bonus (absolute number only).

Solve this problem with the use of one element groups and then solve it again with properly defined groups.

I. Solution with the use of one – element groups.

(a) Table of frequencies for all the employees, regardless their gender:

Items Produced x_i	**Tally**	**Absolute Frequencies** v_i	**Relative Frequencies** $f_i\%$	**Cumulative Absolute Frequencies** N_i	**Cumulative Relative Frequencies** $F_i\%$
1					
2					
3					
4					
5					
6					
7					
8					
9					
10					
	Total				

(b) employees who failed to produce 5 or more items:

(c) employees who produced 5 or more items:

(d) employees who won a bonus:

(e) Table of frequencies for all the employees, where the numbers of men (M) are separated from the numbers of the women (W)

Items Produced x_i	Tally		Absolute Frequencies v_i		Cumulative Absolute Frequencies F_i	
	M	W	M	W	M	W
1						
2						
3						
4						
5						
6						
7						
8						
9						
10						
		Total =				

(f) men who failed to produce 5 or more items:

(g) women who failed to produce 5 or more items:

(h) men who produced 5 or more items:

(i) women who produced 5 or more items:

(j) men who won a bonus:

(k) women who won a bonus:

II. Solution with properly defined groups

(a) Table of frequencies for all the employees, regardless their gender:

Items Produced	Tally	Absolute Frequencies v_i	Relative frequency $f_i\%$	Cumulative Absolute Frequencies N_i
	Total			

(b) employees who failed to produce 5 or more items:

(c) employees who produced 5 or more items:

(d) employees who won a bonus:

(e) Table of frequencies for all the employees, where the numbers of men (M) are separated from the numbers of the women (W)

Items Produced	Tally		Absolute Frequencies v_i		Cumulative Frequencies F_i	
	M	W	M	W	M	W
		Total =				

(f) men who failed to produce 5 or more items:

(g) women who failed to produce 5 or more items:

(h) men who produced 5 or more items:

(i) women who produced 5 or more items:

(j) men who won a bonus:

(k) women who won a bonus:

EXAM PRACTICE QUESTION 6.24

(ANSWERS: mean = 39.0 g to 1 d.p., median class = 31 – 40, modal class = 31 – 40)

Find an estimate for the mean, median class and modal class of the mass of the quail's eggs below:

Mass (g)	Frequency (v_i)
1 – 20	1
21 – 30	3
31 – 40	12
41 – 50	9
61 – 70	2

Mean value =

Median class =

Modal class =

EXAM PRACTICE QUESTION 6.25

(ANSWERS: (a) 15.45, (b) 15, (c) $Q_1 = 14$, $Q_3 = 17$, (d) 3, (e) 1, 2, 6, 8, 13, 21, 27, 31, 34, 39, 40, (f) 32.5%, (g) 15%, (h) 20%)

The values that were measured during an experiment were grouped in classes as shown in the following table, where the classes are represented by their central values:

Central value x_i	Absolute Frequency v_i	Product $v_i x_i$
10	1	$10 \times 1 = 10$
11	1	$11 \times 1 = 11$
12	4	$12 \times 4 = 48$
13	2	$13 \times 2 = 26$
14	5	$14 \times 5 = 70$
15	8	$15 \times 8 = 120$
16	6	$16 \times 6 = 96$
17	4	$17 \times 4 = 68$
18	3	$18 \times 3 = 54$
19	5	$19 \times 5 = 95$
20	1	$20 \times 1 = 20$
Total	$\sum v_i = 40$	$\sum v_i x_i = 618$

Find:

(a) the mean value of the data shown in the table:

(b) The median:

(c) The lower and upper quartiles (Q_1 and Q_3):

(d) The interquartile range:

(e) The cumulative frequencies, *Ni*.

Central value x_i	Absolute Frequency v_i	Product $v_i x_i$	Cumulative Absolute Frequency N_i
10	1	$10\times1=10$	
11	1	$11\times1=11$	
12	4	$12\times4=48$	
13	2	$13\times2=26$	
14	5	$14\times5=70$	
15	8	$15\times8=120$	
16	6	$16\times6=96$	
17	4	$17\times4=68$	
18	3	$18\times3=54$	
19	5	$19\times5=95$	
20	1	$20\times1=20$	
Total	$\sum v_i=40$	$\sum v_i x_i=618$	

(f) Which is the percentage of values less than 15?

(g) Which is the percentage of values more than 18?

(f) Which is the percentage of values equal to 15?

Topic 6.5 GRAPHICAL REPRESENTATION OF THE STATISTICAL DATA.

Bar Charts

EXAM PRACTICE QUESTION 6.26

The following table gives the distribution of frequencies for the activities during the weekends of the employees of a company. In the given space draw the bar chart of the frequencies for these activities.

i	**Activities (x_i)**	**Absolute Frequency v_i**
1	Computers	3
2	Athletics	4
3	Meeting with friends	3
4	Listening to music	7
5	Going to the cinema	7
6	Reading Books	2
7	Other	2

EXAM PRACTICE QUESTION 6.27 (ANSWER: Black)

Jeremy asked 27 of his school mates to tell him their favourite colours for cars. The results are shown in the table below.

(a) Which colour did most of his school mates say?

(b) In the given space draw a bar chart to show his results.

Favourite Colours	**Tally**	**Frequency v_i**
Black	𝍸 𝍸 ||	12
White	𝍸 |	6
Green	||||	4
Blue	|||	3
Red	||	2

Line Charts

EXAM PRACTICE QUESTION 6.28

30 people were asked to indicate the number of people they live with. Their responses are presented in the following table:

Family status	Number of people that live with me x_i	Absolute Frequency v_i
single	0	7
married – no children	1	4
married – one child	2	12
married – two children	3	5
married – three children	4	2
	Total =	30

In the given space, draw the line diagram for the above data

Histogram of Frequencies

EXAM PRACTICE QUESTION 6.29

Equal Class Width

The structural engineers want to calculate the quality of the finished surface of the entrance passage to a parking area which will be used only by private cars (no heavy lorries, tracks and busses will be allowed). Depending on the number of passengers, the manufacture and the additional load the cars will carry, their weight varies. So, they made an experiment on 30 different cars and they collected the following results:

Weight of the car in Kg	Number of cars (absolute frequency) v_i
$500 \le x < 700$	5
$700 \le x < 900$	7
$900 \le x < 1\,100$	11
$1\,100 \le x < 1\,300$	2
$1\,300 \le x < 1\,500$	5
Total =	30

Draw the Histogram of Frequencies of this experiment.

Weight of the car in Kg	Central value x_i	Number of cars (absolute frequency) v_i	Frequency Density
$500 \le x < 700$		5	
$700 \le x < 900$		7	
$900 \le x < 1\,100$		11	
$1\,100 \le x < 1\,300$		2	
$1\,300 \le x < 1\,500$		5	
	Total =	30	

EXAM PRACTICE QUESTION 6.30

Unequal Class Widths

The durations of telephone calls for 120 people in 1 day were measured and they were recorded as follows:

Duration of a telephone call in sec	Number calls (absolute frequency) v_i
$0 \le t < 15$	20
$15 \le t < 35$	25
$35 \le t < 40$	30
$40 \le t < 55$	45
Total =	120

Draw the corresponding histogram of frequencies.

Duration of a telephone call, in sec	Central value t_i	Class Width in sec	Number calls (frequency) v_i	Frequency Density
$0 \le t < 15$			20	
$15 \le t < 35$			25	
$35 \le t < 40$			30	
$40 \le t < 55$			45	
		Total =	100	

Pie Chart

EXAM PRACTICE QUESTION 6.31 **(ANSWERS: number of students with grade 9 = 45 or 10%, grade 8 = 135 or 30%, grade 7 = 180 or 40%, grade 6 = 90 or 20%)**

In a sample of 450 students of years 10 and 11 whose grades were 9, 8, 7 and 6 it was observed that 30% of the students obtained grade 8 and that the students with grade 6 were twice as many as those with grade 9.

Fill in the given table and then draw the pie chart of this sample, when you know that the angle of the sector corresponding to the students who got 7 has central angle 144°. Then, determine the number of students with grades 9, 8, 7 and 6 and the corresponding percentages.

Also, draw the bar chart for this sample.

Grade x_i	**Frequency (Number of students) v_i**	**Angle**	**Relative Frequency %**
9			
8			
7			
6			
Total			

The pie chart is:

The Bar Chart is:

EXAM PRACTICE QUESTION 6.32 **(ANSWERS: (a) the Food department, (b) Total amount spent altogether £ 3 673 469.39 (c) Food £ 1 040 816.33, Clothes & Shoes £ 693 877.55, Toys £ 561 244.49, Furniture £ 377 551.02)**

A Department store made a statistical analysis on the money their customers spend on Food, Clothes & Shoes, Toys, Furniture and Petrol. The results for the month September are presented in the table and pie chart of the following figure.

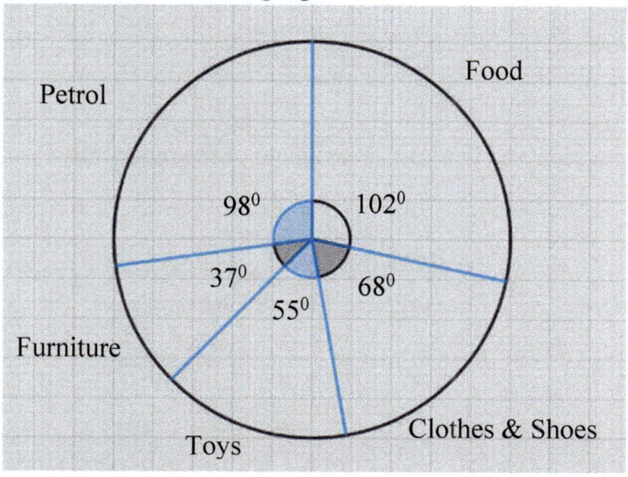

Department x_i	**Angle**	**Money Spent**
Food	102°	
Clothes & Shoes	68°	
Toys	55°	
Furniture	37°	
Petrol	98°	£1 000 000.00

If the customers spent £1 000 000 on petrol, calculate:

(a) In which department did the customers spent most money?

(b) How much was the total amount spent altogether?

(c) How much money was spent on each individual department and

(d) fill in the above table.

Frequency Polygon

EXAM PRACTICE QUESTION 6.33 (ANSWER: 38 grams)

Unequal class widths

The following table shows information about the mass in grams, of 100 chicken eggs.

Mass of chicken Eggs in grams	Frequency v_i
$20 < m \leq 30$	20
$30 < m \leq 40$	30
$40 < m \leq 45$	25
$45 < m \leq 50$	25

Draw a frequency polygon to represent this information and workout an estimate of the mean mass.

egg mass in grams	central value m_i	Absolute Frequency v_i	Product $v_i m_i$
$20 < m \leq 30$		20	
$30 < m \leq 40$		30	
$40 < m \leq 45$		25	
$45 < m \leq 50$		25	
	Total =	100	

EXAM PRACTICE QUESTION 6.34 (ANSWER: 27.3 cm)

Equal class widths

The adjacent table shows some information about the heights of 50 toys.

Draw a frequency polygon to represent this information and workout an estimate of their mean height.

Height (h) cm	Frequency v_i
$15 < h \leq 20$	5
$20 < h \leq 25$	8
$25 < h \leq 30$	15
$30 < h \leq 35$	12
$35 < h \leq 40$	10

Height (h) cm	central value h_i	Frequency v_i	Product `$h_i v_i$
$15 < h \leq 20$		5	
$20 < h \leq 25$		8	
$25 < h \leq 30$		15	
$30 < h \leq 35$		12	
$35 < h \leq 40$		10	
	Total		

EXAM PRACTICE QUESTION 6.35 (ANSWER: 11.49 seconds to 2 d.p.)

Time (t) seconds	Frequency v_i
$11 < t \leq 11.2$	5
$11.2 < t \leq 11.4$	8
$11.4 < t \leq 11.6$	9
$11.6 < t \leq 11.8$	7
$11.8 < t \leq 12.0$	5

34 male students ran a 100 m sprint. Each student's time was record.
The adjacent table shows information about their times:
Draw a frequency polygon to represent this information and workout an estimate of the mean time to 2 d.p.

Time (t) seconds	Central value (t_i) sec	Freq. v_i	product $t_i v_i$
$11 < t \leq 11.2$		5	
$11.2 < t \leq 11.4$		8	
$11.4 < t \leq 11.6$		9	
$11.6 < t \leq 11.8$		7	
$11.8 < t \leq 12.0$		5	

EXAM PRACTICE QUESTION 6.36

The following table shows the frequency distribution of marks in a chemistry test in two different classes:

Marks (M)	Central value M_i	Class A	Class B
$0 < M \leq 50$	25	5	1
$50 < M \leq 60$	55	8	8
$60 < M \leq 70$	65	9	7
$70 < M \leq 80$	75	7	8
$80 < M \leq 100$	90	5	6
	Total =	30	30

Draw the frequency polygons for the two classes on the same grid. Compare the two classes.

Cumulative Histograms or Ogives

EXAM PRACTICE QUESTION 6.37

(ANSWERS: (b) $Q_2 = 33$ min, (c) $Q_1 = 23$ min, $Q_3 = 42$ mm $IQR = 19$ min, (d) 20%)

The following table shows a summary of the time it took 100 players to complete a Puzzle.

Time t (minutes)	Upper Class Boundary	Frequency v_i	Cumulative Frequency
$0 < t \leq 10$	10	5	5
$10 < t \leq 20$	20	15	20
$20 < t \leq 30$	30	20	40
$30 < t \leq 40$	40	30	70
$40 < t \leq 50$	50	20	90
$50 < t \leq 60$	60	10	100

(a) Draw a cumulative frequency graph for the data.
(b) Estimate the median for this data.
(c) Estimate the lower and upper quartiles and hence the interquartile range for this data.
(d) Estimate the percentage of players that will complete the puzzle in more than 45 minutes.

EXAM PRACTICE QUESTION 6.38

(ANSWERS: $Q_2 = 43$ s, $Q_1 = 33$ s, $Q_3 = 52$ s, $IQR = 19$ s)

Time taken to eat 2 eggs (seconds)	Frequency
$10 < t \leq 20$	3
$20 < t \leq 30$	7
$30 < t \leq 40$	12
$40 < t \leq 50$	18
$50 < t \leq 60$	11
$60 < t \leq 70$	5

Use a cumulative frequency graph to find an estimate for the median, the quartiles and the interquartile range for the data presented in the adjacent table:

Time taken to eat 2 eggs (seconds)	Upper Class Boundary	Frequency	Cumulative Frequency
$10 < t \leq 20$		3	
$20 < t \leq 30$		7	
$30 < t \leq 40$		12	
$40 < t \leq 50$		18	
$50 < t \leq 60$		11	
$60 < t \leq 70$		5	

Box and Whisker Plots

Exam Practice Question 6.39

Fifty pupils sat a Mathematics examination. The lowest mark gained was 35 and the highest was 95. The quartiles of the marks were 55, 70, 80 respectively.
Draw a box plot to represent these data.

EXAM PRACTICE QUESTION 6.40

Fifty boys of year 11 took part in a Mathematics test. The box plot below shows the information about the marks achieved by the boys. The test is marked out of 100.

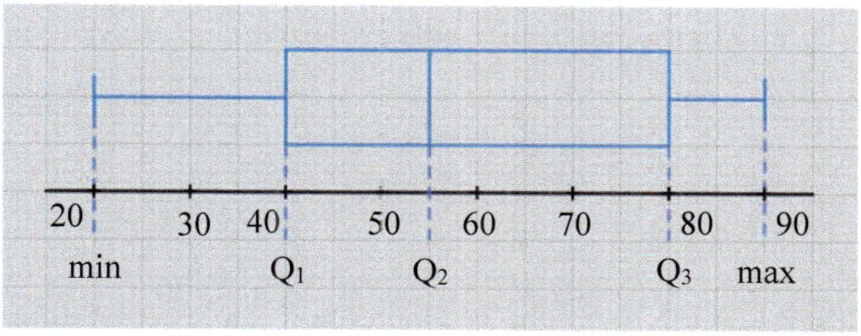

Fifty girls of year 11 took part in a same test. The achieved results were: The lowest mark was 25 and the height was 98. The quartiles of the marks were 50, 65, 80 respectively.

In the given space redraw the above box plot and add a box plot to represent the data for the girls. Compare the results with those of the boys.

EXAM PRACTICE QUESTION 6.41

Pupils in year 11 are all given the same test. The test is marked out of 100.
The results for the boys is shown in the following figure and for the girls are: Lowest score 14 marks, Highest score 100 marks, Lower quartile 40 marks, Median 60 marks, Upper quartile 80 marks.

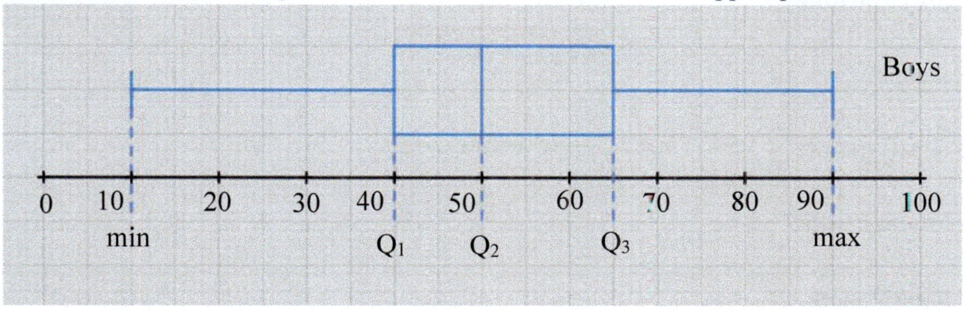

(a) In the given space redraw the above box plot and add the box plot for the girls' results below the number line.

(b) Comment on the difference between the boys' and the girls' results.

Exam Practice Question 6.42

The marks of a test in maths (marked out of 100) for boys and girls are shown on the following box plots.

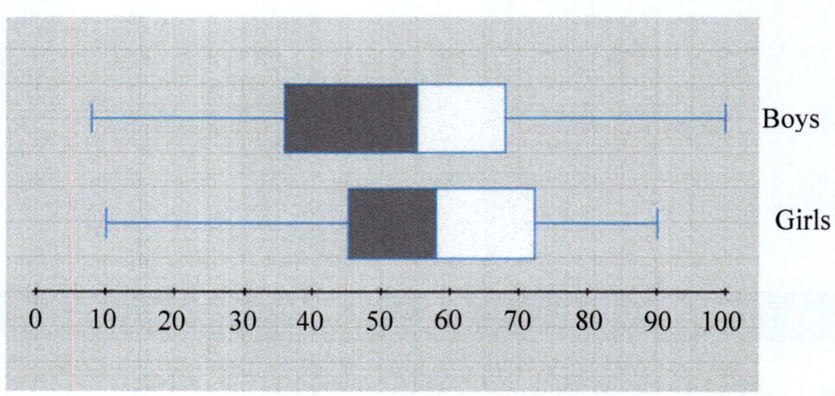

State for each (Boys and Girls):

(i) the lower quartile

(ii) the median

(iii) the upper quartile

(iv) the lowest mark

(v) the highest mark

Topic 6.6 STEM AND LEAF DIAGRAMS.

EXAM PRACTICE QUESTION 6.43

Consider the following list of 30 numbers:

55	24	62	31	64
25	33	20	31	41
33	32	22	44	55
46	45	65	43	44
56	60	55	23	21
66	21	21	42	32

(a) Construct an unordered stem-and-leaf diagram.

(b) Then, construct an ordered stem-and-leaf diagram.

Exam Practice Question 6.44

The marks (out of 100) of a Mathematics test of 40 students are recorded in the adjacent table:

47	25	64	44	52	44	42	90	100	55
98	54	75	39	55	43	75	60	90	60
12	47	52	42	79	25	85	55	30	58
39	42	55	60	23	79	50	40	60	55

Construct a stem-and-leaf diagram to represent them: (a) unordered (b) ordered (c) splitting the stem.

EXAM PRACTICE QUESTION 6.45

The adjacent table shows the ages of 15 adults:
Show this information in an ordered stem and leaf diagram.

35	22	45	33	27
54	29	29	35	37
55	57	47	48	46

EXAM PRACTICE QUESTION 6.46 (ANSWER: (b) 32)

50 people in a buss stop were asked how many minutes they are waiting for the next bus to arrive.
The results are shown below:

10	44	17	38	65	10	45	25	10	15
45	39	29	36	50	15	58	54	56	59
45	23	35	40	49	39	46	57	50	54
22	45	40	47	44	56	58	46	58	60
44	16	45	45	47	54	53	48	68	65

(a) Construct a stem-and-leaf diagram to represent these answers.

(b) How many people waited more than 40 minutes?

EXAM PRACTICE QUESTION 6.47 (ANSWERS: (a) 327.5 mm (b) 352 mm)

The heights, in millimetres of 20 plants were measured and the corresponding stem and leaf diagram is shown in the adjacent table.

(a) Find the Median.
(b) Workout the Range.

Heights key: 1|45 means 145 mm

Stem	**Leaf**						
1	45	45	47				(3)
2	5	15	33	37	50		(5)
3	25	27	28	28	28	28	(6)
4	23	25	27	45	95	97	(6)
							20

EXAM PRACTICE QUESTION 6.48 (ANSWER: (b) P(less than 45) = $\frac{1}{3}$)

59	58	57	55	55
45	46	47	48	48
39	37	38	39	40

The adjacent table presents the ages, in years of 15 lecturers.

(a) Draw an ordered-stem-and-leaf diagram to show this information, including a Key.
(b) One of these lectures is picked at random. Work out the probability that this lecturer is less than 45 years old.

EXAM PRACTICE QUESTION 6.49 (ANSWER: (b) 12 min)

Here are the times, in minutes, taken to solve a mathematics problem.

8	8	10	25	24	9	9	12	13	20
12	9	11	14	15	10	12	14	15	22

(a) Draw a stem and leaf diagram.

(b) Work out the median time to solve this problem

EXAM PRACTICE QUESTION 6.50 (ANSWER: (b) 6 s)

Jack measured the times in seconds it took 25 athletes to run 400 m and produced the adjacent table.

52	58	51	52	55
55	60	61	59	59
54	57	57	49	49
53	55	56	55	55
60	61	61	53	55

(a) Show this information in an ordered stem and leaf diagram.

(b) Workout the Interquartile Range (*IQR*).

EXAM PRACTICE QUESTION 6.51 (ANSWER: (b) 11.1 s)

In the same games as above, Jack also measured the times in seconds it took 25 athletes to run 100 m.

10.8	11.2	10.8	10.9	10.9	11.3	10.9	11.2	11.4	11.5
10.9	11.1	11.1	11.8	11.1	12.0	11.0	11.0	10.9	11.2

(a) Draw an ordered stem and leaf diagram to show this information.
(b) Workout the median.

EXAM PRACTICE QUESTION 6.52 (ANSWERS: (b) min time = 53 min, Median = 64 min, Q_1 = 59 min, Q_3 = 72 min, IQR = 13 min, 6.45%)

A motor Company records the time, t in minutes to the nearest minute, to carry out the full service on a particular model of a car.
The values of t recorded for 31 full services are given in the table below.

(a) Draw an unordered stem-and-leaf diagram and then an ordered one. Label carefully the number of leaves and the key used.

Also,

(b) determine the minimum time measured, the median time and the interquartile range
(c) Work out the probability that a car stays longer than 75 minutes.

59	74	69	60	65	75	72	62	63	74
72	62	55	72	59	57	64	72	65	58
54	59	75	64	79	75	62	85	65	53
59									

EXAM PRACTICE QUESTION 6.53 (ANSWER: 6 cars)

The speeds, in miles per hour, of 15 cars, measured in a built-up area are recorded to the nearest mph.

35	22	31	33	29
37	29	28	35	25
20	23	26	23	34

Draw an unordered and then an ordered stem and leaf diagram. Also, draw a splitting the stem diagram. Label carefully the number of leaves and the key used.

If the speed limit is 30 mph, find the number of cars that are exceeding this limit.

Back to back stem and leaf diagram

EXAM PRACTICE QUESTION 6.54 **(ANSWERS: (b) men 1.33 m, women 1.705 m (c) men: 8.33 m, women: 6.89 m, (d) men: 8.90 m, women: 7.40 m)**

The host cities for the Summer Olympics and the Gold medalists in the Long Jump are as follows:

Men	Length (m)	Women	Length (m)
I – Athens, 1896			
Ellery Clark United States	6.35	-	-
II – Paris, 1900			
Alvin Kraenzlein United States	7.185	-	-
III – St. Luis, 1904			
Myer Prinstein United States	7.34	-	-
IV – London, 1908			
Frank Irons United States	7.48	-	-
V – Stockholm, 1912			
Albert Gutterson United States	7.60	-	-
VI – Berlin 1916 – Cancelled due to WWI			
VII - Antwerp, 1920			
William Petersson Sweden	6.94	-	-
VIII – Paris, 1924			
DeHart Hubbard United States	7.12	-	-
IX – Amsterdam, 1928			
Ed Hamm United States	7.73	-	-
X – Los Angeles, 1932			
Ed Gordon United States	7.64	-	-
XI – Berlin, 1936			
Jesse Owens United States	8.06	-	-
XII – Tokyo, 1940 – Cancelled due to WWII			
XIII – London, 1944 – Cancelled due to WWII			
XIV – London, 1948			
Willie Steele United States	7.825	Olga Gyarmati Hungary	5.695
XV – 1952 Helsinki			
Jerome Biffle United States	7.57	Yvette Williams New Zealand	6.24
XVI – 1956 Melbourne & Stockholm			
Gregory Bell United States	7.83	Elżbieta Krzesińska Polland	6.35

Men	Length (m)	Women	Length (m)
XVII – Rome, 1960			
Ralph Boston United States	8.12	Vera Kolashnikova Soviet Union	6.37
XVIII – 1964 Tokyo			
Lynn Davies Great Britain	8.07	Mary Rand Great Britain	6.76
XIX – 1968 Mexico City			
Bob Beamon United States	8.90	Viorica Viscopoleanu Romania	6.82
XX – 1972 Munich			
Randy Williams United States	8.24	Heide Rosendahl West Germany	6.78
XXI – 1976 Montreal			
Arnie Robinson United States	8.35	Angela Voigt East Germany	6.72
XXII – 1980 Moscow			
Lutz Dombrowski East Germany	8.54	Tatyana Kolpakova Soviet Union	7.06
XXIII – 1984 Los Angeles			
Carl Lewis United States	8.54	Anişoara Stanciu Romania	6.96
XXIV – 1988 Seoul			
Carl Lewis United States	8.72	Jackie Joyner-Kersee United States	7.40
XXV – 1992 Barcelona			
Carl Lewis United States	8.67	Heike Drechsler Germany	7.14
XXVI – 1996 Atlanta			
Carl Lewis United States	8.50	Chioma Ajunwa Nigeria	7.12
XXVII – 2000 Sydney			
Iván Pedroso Cuba	8.32	Fiona May Italy	6.81
XXVIII – 2004 Athens			
Dwight Phillips United States	8.59	Tatyana Lebedeva Russia	7.07
XXIX – 2008 Beijing			
Irving Saladino Panama	8.34	Maurren Higa Maggi Brazil	7.04
XXX – 2012 London			
Greg Rutherford Great Britain	8.31	Brittney Reese United States	7.12
XXXI – 2016 Rio De Janeiro			
Jeff Henderson United States	8.38	Tianna Bartoletta United States	7.17